METABOLOMICS - FUNDAMENTALS AND APPLICATIONS

Edited by **Jeevan K. Prasain**

Metabolomics - Fundamentals and Applications
http://dx.doi.org/10.5772/62614
Edited by Jeevan K. Prasain

Contributors
Francisco Javier Toledo Marante, Rangaswamy Lakshminarayana, Bangalore Prabhashankar Arathi, Poorigali Raghavendra-Rao Sowmya, Kariyappa Vijay, Vallikannan Baskaran, Josep Rubert, Laura Righetti, Chiara Dall'Asta, Jana Hajslova, Ana Colette Maurício, Gayla Ivanova, Tiago Pereira, Rita Caseiro, Pétia Georgieva, Jianxin Chen, Michele Nishiguchi, Stephen Grace, Dane Hudson, Marcos Soto-Hernández, Emmanuel Ibarra-Estrada, Mariana Palma Tenango, Qifang Pan, Natali Rianika Mustafa, Robert Verpoorte, Kexuan Tang, Luis Aldámiz-Echevarría, Fernando Andrade, Domingo González-Lamuño, Marta Llarena, Mina Hanna, Patrick D. Brophy

Published by InTech
Janeza Trdine 9, 51000 Rijeka, Croatia

Publishing Process Manager Dajana Pemac

Technical Editor SPi Global

Cover InTech Design team

First published January, 2017

Printed in Croatia

Additional hard copies can be obtained from orders@intechopen.com

Metabolomics - Fundamentals and Applications, Edited by Jeevan K. Prasain

p. cm.

Print ISBN 978-953-51-2853-3
Online ISBN 978-953-51-2854-0

For EU product safety concerns:
IN TECH d.o.o., Prolaz Marije Krucifikse Kozulić 3, 51000 Rijeka, Croatia,
info@intechopen.com or visit our website at intechopen.com.

Meet the editor

Dr. Jeevan K. Prasain received his PhD in 1998 from the Toyama Medical and Pharmaceutical University, Japan, and then joined the Biotechnology Research Institute, Montreal, Canada, as a visiting fellow where he contributed to the development of high-throughput screening of bioactive natural products and their analysis by tandem mass spectrometry. He was a postdoctoral fellow in 2000 at the Institute of Physical and Chemical Research (RIKEN), Japan. Currently, Dr. Prasain is an assistant professor at the Department of Pharmacology & Toxicology, University of Alabama at Birmingham, and his research interests include metabolomics/lipidomics, biological evaluation of dietary supplements in vitro and in vivo, metabolism, bioavailability, and pharmacokinetic evaluation of active metabolites.

Contents

Preface

Metabolomics being system-level analyses of a large number of metabolites in response to biological stimuli or environmental perturbations is an emerging powerful technique with a wide range of applications including clinical chemistry. Over the next decade, metabolomics is expected to provide more insights into quality of foods, disease onset, and progression and identification of biomarkers and thus may revolutionize the future healthcare strategy. Given its potential applications, there is still a dearth of books that capture the current status of our knowledge in various aspects of metabolomics. This book is an attempt to understand basic principles and applications of metabolomics. It contains 11 chapters divided into three main sections covering basic principles, analytical techniques, data analysis and applications in foods, plant metabolism, agriculture, and biomedical and clinical sciences.

Thanks are expressed to the contributing authors who have attempted to provide updated review from the basic principles to practical application of metabolomics. Finally, I would like to thank Ms. Dajana Pemac, Publishing Process Manager, and other InTech publication staffs for their valuable support.

Jeevan K. Prasain, PhD
Department of Pharmacology & Toxicology
University of Alabama at Birmingham
USA

Section 1

Analysis and Data Processing

Chapter 1

Marine Environmental Metabolomics

Francisco Javier Toledo Marante

Additional information is available at the end of the chapter

http://dx.doi.org/10.5772/66246

Abstract

Marine environmental metabolomics studies the interactions of marine organisms with their environment using metabolomics to characterise these interactions. There are many advantages in using this method to study interactions between organisms and the environment and to assess the function and health of organisms at the molecular level. In fact, metabolomics is finding an increasing number of applications in the marine sciences. These range from understanding the response of organisms to abiotic pressure to researching the response of organisms to other biota. These interactions can be studied at different levels, from individuals to populations for more traditional eco-physiological or ecological studies. Marine organisms have developed a high diversity of chemical defences to avoid predators and parasites. This study therefore highlights the complexity of chemical interactions in the marine environment. The research methods include ^{1}H- and ^{13}C-NMR spectroscopy, mass spectrometry, analytical and preparative chromatography, and a multitude of bio-assays.

Keywords: metabolomics, marine, environmental, allomones, kairomones, pheromones

1. Introduction

Environmental metabolomics is the application of metabolomics for characterising the interactions of organisms with their environment [1]. Marine environmental metabolomics is the application of metabolomics to characterise the interactions of marine organisms with their environment [2].

As Greek scholars claimed, everything is born out of struggle and need. The pressing need of organisms is to adapt to the environment or adapt the environment to the interests of the species. The survival and progress of the different living entities represent a secret driving force that harnesses metabolites to trigger sophisticated chemical interactions. Although the production of these bio-active substances requires an enormous effort from the organism in terms of energy, the adaptive advantage gained in return is as spectacular as it is necessary for survival.

The progressive degradation of the marine environment on the other hand (animal/plant pests, pollution, etc.) leads us to the need to protect it, and to do so, we have to understand it. Scientists are increasingly clear that our understanding of the marine environment is incomplete without a deeper understanding of its metabolomics. There is even talk of a new science, although its name is not clear. Some call it chemical ecology [3], others ecological bio-chemistry [4], and others consider it part of synecology or biocenotics. According to the principle of science itself "given enough time, only the necessary will survive" [3] so one can expect that the day will come when chemists and biologists will decide on the best name for it.

2. Terminology

Coactones or semiochemicals are the compounds released by an organism which evoke a reaction in another organism of a different or the same species. When these compounds act at a distance, they are called allelochemicals, and their interaction is known as allelopathy [5].

Interactions can be inter-specific or intra-specific, depending on whether they affect individuals of different or the same species, respectively. The chemical factors that affect organisms of different species can in turn be allomones or kairomones, while the chemical factors that affect individuals of the same species can be by auto-toxins, or pheromones [3].

Allomones are semiochemicals that favour the emitter, but not for the receiver. Examples in the marine environment include toxins, digestibility reducing factors, repellents, feeding deterrents, anti-fouling compounds, escape substances, suppressors—antibiotics and cytotoxins [6].

Kairomones are semiochemicals that favour the receiver. Examples from the marine environment include predator attractors or substances that predators use to locate their prey, adaptation inducers, like the spine-development factor in rotifers, warnings signals of danger or toxicity, which benefit the receiver, such as colourants that generate bright colours with characteristic designs on the most toxic animals, growth stimulators, etc. [6].

Pheromones are the semiochemicals released into the environment to influence behaviour or some biological function in the same species. These include sexual/social/warning/territorial marking/trace and communication pheromones [7]. One particular and highly important case of the latter refers to migration pheromones or tele-mediators [8].

3. Toxins

Animals that are mobile or have hard shells or spines are typically not defended by noxious or toxic chemicals. This is the case of the sea urchin or the spiny lobster. Contrarily, the spotted trunkfish (*Lactophrys bicaudalis*) secretes a colourless toxin from glands on its skin when touched. Predators as large as nurse sharks can die as a result of eating a trunkfish. Methane chemical ionisation gas chromatography-mass spectrometry was used to study pahutoxin (**1**) and choline chloride esters of 16C (**2**), 17C (**3**) and 18C (**4**) fatty acids from Caribbean trunkfish (*Lactophrys triqueter*) toxin (**Figure 1**) [9].

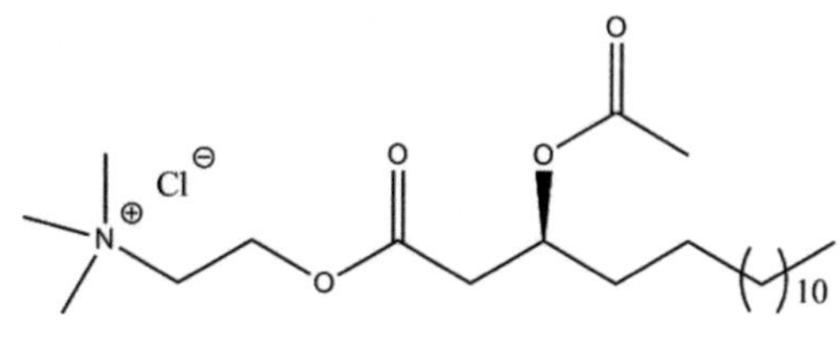

1

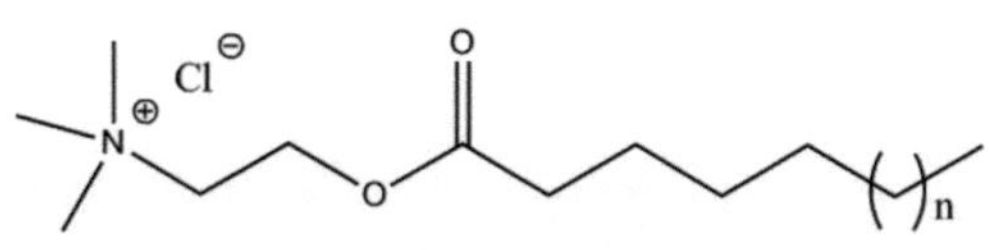

2, n=10
3, n=11
4, n=12

Figure 1. Caribbean trunkfish toxins.

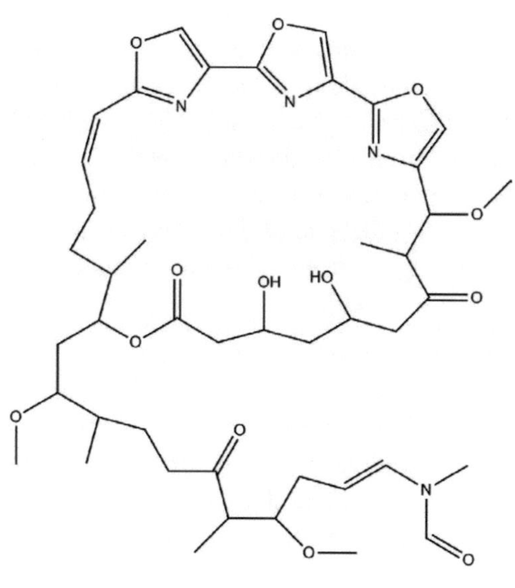

5

Figure 2. Jaspisamide A, a nudibranch egg ribbon toxin.

On the other hand, marine snails are strongly protected. Nudibranchs, or sea slugs as they are also known, are an example that is commonly quoted of how organisms with powerful chemical defences have little need for a physical defence like a hard shell to protect them from predators. Nudibranchs usually obtain their chemical defences from the sponges, bryozoans and sea squirts that they eat, but cases have been reported in which these are produced by *de novo* bio-genesis [10]. Nudibranchs also put defensive compounds in their soft egg ribbons. One known example is the case of the jaspisamides A (**5**), B and C, chemical compounds that, while produced by the Okinawan sponge *Jaspis* sp., have been isolated from nudibranch egg masses (**Figure 2**) [11]. These trisoxazole macrolides are cytotoxic and antifungal metabolites initially isolated from the egg ribbons of the *Hexabranchus* nudibranch [12]. They possess a characteristic macrolide portion, comprising three contiguous oxazole units. Trisoxazole macrolides depolymerise F-actin and form a 1:1 complex with G-actin, thereby exhibiting potent toxicity towards eukaryotic cells [13].

4. Repellents/feeding deterrents

Sponges are an abundant group of coral-reef invertebrates that are very chemically rich. Recent studies have shown that many sponge chemicals effectively repel potential predators, and many of the distasteful compounds have now been isolated and structurally characterised [14]. Bioassays are usually run to locate sponges that accumulate repellents. As an experimental methodology, *Preference assays* offer a range of potential prey species to common predators. Species that are avoided by predators frequently have chemical defences. *Caging experiments* also cast light on the role played by predators in habitats, such as coral reefs, where they abound, in eliminating poorly defended prey. Ecologists first identify low preference prey species, before determining whether they are equipped with a chemical defence by adding chemical extracts taken from them to a food item that predators readily eat. This feeding experiment is simple: each fish (from a typical sample of 10–15 individuals) is offered a food pellet containing the extract and an identical food pellet without the extract. The numbers of control and treatment pellets eaten are recorded in tables and graphs and then compared to see whether the fish find the extract distasteful. The assay is then made more realistic from an ecological point of view by placing control and extract-treated foods on the reef where many different species of fish can feed on them.

Soft fleshy seaweeds found where herbivorous fish and invertebrates abound typically deter herbivory by producing distasteful secondary metabolites. *Halimeda* spp. are among the most common seaweeds on tropical reefs where herbivory is intense. These calcified seaweeds produce feeding deterrents. Halimedatrial (**6**) a structurally unprecedented diterpenoid trialdehyde was identified as the major secondary metabolite in six species of these calcareous reef-building alga (**Figure 3**) [15]. In laboratory bioassays, this metabolite is toxic to reef fish, significantly reducing feeding in herbivorous fishes, and has cytotoxic and anti-microbial activities. When plants from most of the *Halimeda* spp. on Guam suffer grinding or crushing damage, they immediately convert halimedatetraacetate, a less-deterrent secondary metabolite, into halimedatrial, a more powerful deterrent to feeding (**6**). The conversion process would be triggered when fish bite or chew *Halimeda* plants. The process of rapid conversion is known as activation. Extracts from injured plants contained more halimedatrial and were more deterrent towards herbivorous fishes than extracts from control plants. Herbivore-activated defences are common in many families of terrestrial plants, but this was the first example of an activated defence in a marine plant [16].

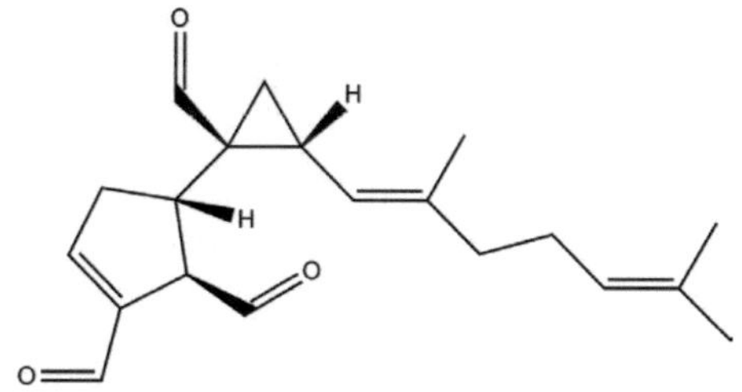

6

Figure 3. Halimedatrial.

Gorgonians, a type of soft coral, are close relatives of hard corals, but they do not have a hard calcium carbonate skeleton. Their soft texture seems to make them a target for a range of reef predators, but the many novel compounds they produce act as an effective defence to protect them from these predators. Hence, for example, *Erythropodium caribaeorum* and *Verrucella umbraculum* produce several diterpenes B, **7** (**Figure 4**) [17–20].

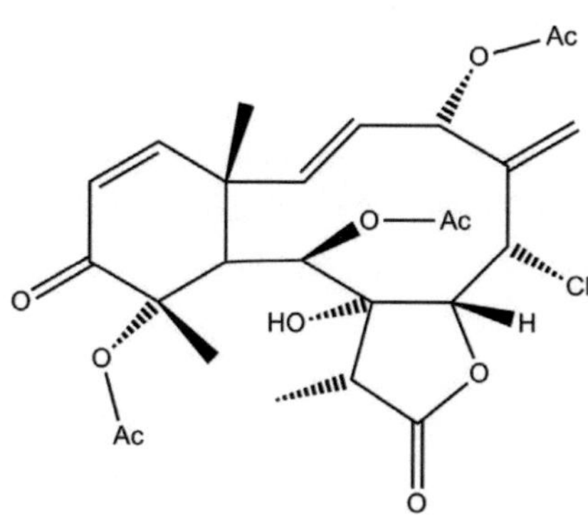

7

Figure 4. Erythrolide B.

Biomass screening of a new marine-derived strain of *Penicillium roqueforti,* as produced by liquid-state fermentation, led to the identification of the compound 4-hydroxy-benzaldehyde **8** (0.92%) [21]. This natural product is a feeding deterrent factor that restrains the greatest predator of the *Isodictya erinacea* sponge, the *Perknaster fuscus* starfish [22]. Although it is the first time that this has been described in fungi, this substance is structurally related to other previously known metabolites in these organisms that are involved in the shikimic acid pathway, such as oxime-2-(4-hydroxyphenyl)-2-oxo acetaldehyde **9**, a metabolite previously isolated from *P. olsonii* (**Figure 5**) [23].

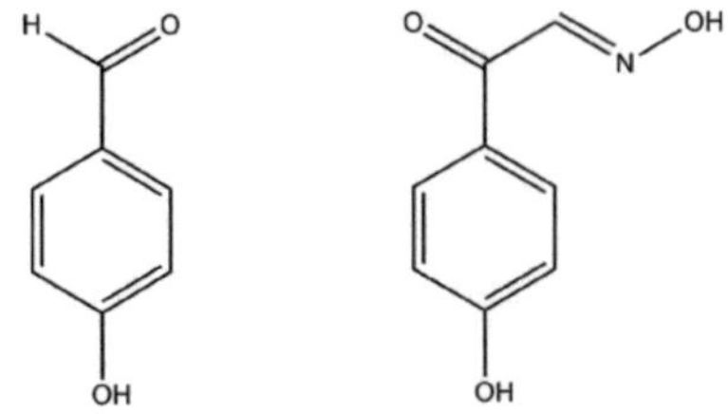

8 9

Figure 5. Allomones of *Penicillium roqueforti* (**8**) and *Penicillium olsonii* (**9**).

This seems to suggest that the shikimic acid pathway allows fungi to produce their allomones by *de novo* bio-genesis. After these and many other analogous discoveries [2], the idea began to take hold that these allomones are not produced by sponges, but by some symbiotic fungus that lives on them. One fact that supports this idea is that other fungal allomones have been identified among the components of others sponges [24], oysters [25] and algae [26, 27].

5. Anti-fouling compounds

Many sessile marine organisms are surprisingly clean given the abundance of algal spores and invertebrate larvae that could settle and grow on them. Some seaweeds and invertebrates produce compounds that deter or kill larvae and spores attempting to colonise them as a way of keeping clean. Zosteric acid **10** (**Figure 6**), isolated from the young shoots of a seagrass, is an example of a potent natural anti-fouling compound [28, 29].

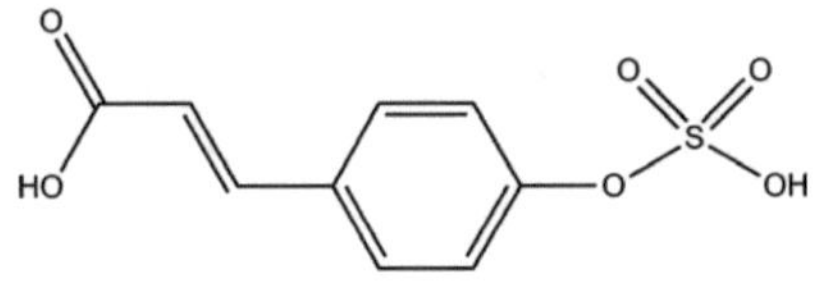

10

Figure 6. Zosteric acid.

6. Kairomones

Saponins act as repellents in sea cucumbers, and many species produce these cytotoxic secondary metabolites. Despite the deterrent, they are still colonised by multiple symbiotic organisms, including the Harlequin crab, *Lissocarcinus orbicularis*, which is one of the most widespread in the Indo-Pacific Ocean. The authors have identified the nature of the molecules secreted by sea cucumbers that attract symbionts for the first time. The kairomones recognised by the crabs are saponins—like holothurin A, **11** (**Figure 7**)—ensuring symbiosis. The success of this symbiosis is due to the ability that crabs showed during evolution to overcome the sea cucumber's chemical defences, with their repellents evolving into powerful attractants [30].

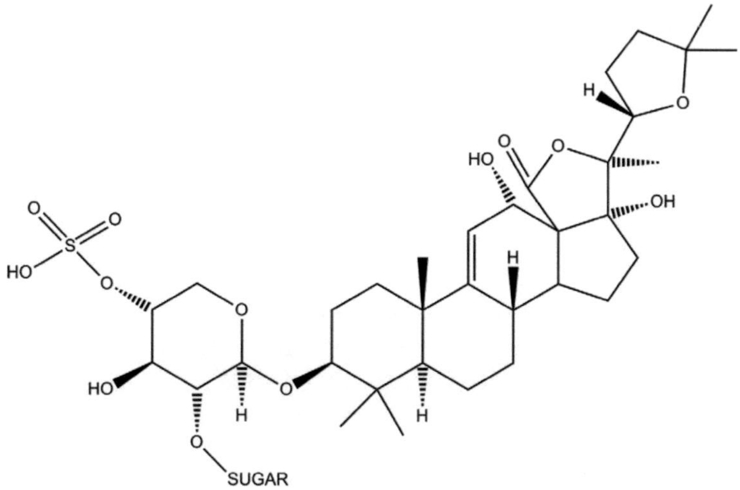

11

Figure 7. Holothurin A.

7. Pheromones

Pheromones are secreted or excreted chemicals that trigger a social response in members of the same species. In the marine environment, there are some well-known examples that affect behaviour or physiology: alarm pheromones, sex pheromones, etc. There are papers in the field of alarm pheromones reporting how the nudibranchs *Tambje* spp. use alarm pheromones,

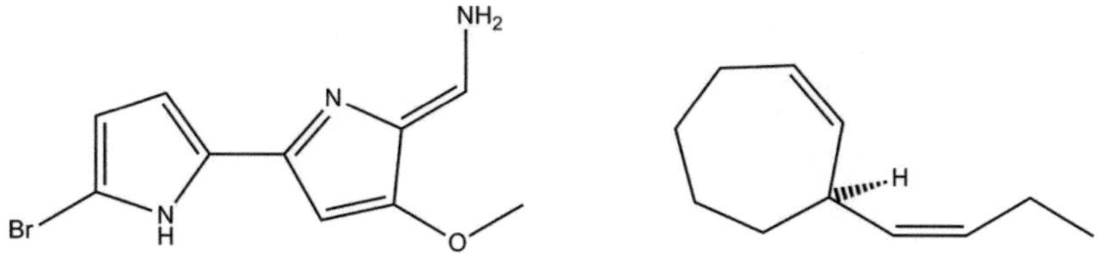

12 **13**

Figure 8. Tambjamine B (**12**) and ectocarpene (**13**).

such as tambjamine B, **12** (**Figure 8**), to alert others to a threat [31]. Similarly, there are papers in the field of sex pheromones that report how male copepods can follow a three-dimensional pheromone trail left by a swimming female [32], and male gametes of many brown algae use a pheromone, such as ectocarpene, **13,** (**Figure 8**) to help find a female gamete for fertilisation, a phenomenon known as chemotaxis [33, 34], in fact there are even videos of this on internet.

8. Research methods

Isolating and identifying natural products requires the use of physio-chemical fractionation and purification techniques (**Figure 9**). These products can be explored once enough biological matter is obtained, either from organisms collected directly from their natural habitat or using bio-processes (fermentation, photo-bio-reaction) or marine aquaculture to grow them. The biomass obtained can be frozen, freeze-dried or chemically set in a dissolvent to conserve it.

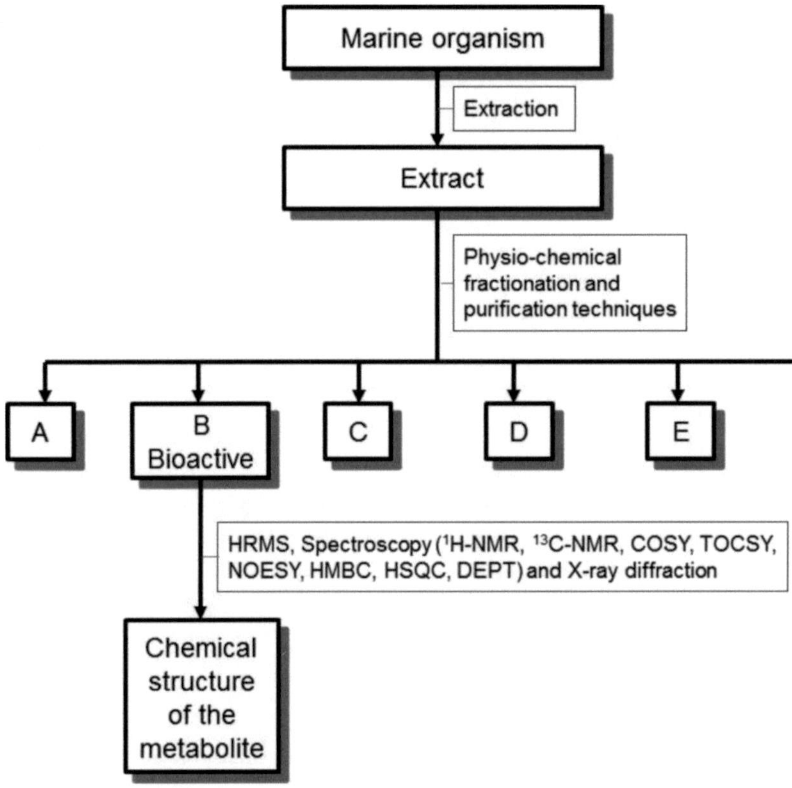

Figure 9. Methodology to identify bioactive substances.

The preliminary separation techniques used in laboratories are performed with adsorption chromatography, using gravity flow columns at low or medium pressure, or using liquid-

liquid partition methods [35]. The latter technique can be applied simply with the help of a decanting funnel, and using one of several variants, it can provide low, medium or high-polarity fractions. This was the case of the fractions obtained from the growth medium used for *P. roqueforti* fungus [36]. Throughout the process to separate the components, there is always the option of using a *Sephadex*-type molecular size exclusion chromatography, which separates families of chemical components of similar molecular size. There are several resins on the market under the common name of the technique (*Gel Filtration*). A very popular one is *LH-20* [37]. Other separation techniques can then be used, such as adsorption chromatography in flash-type columns, either in normal phase or in reverse phase, allowing us to work on a scale of grams, or the case of preparatory plates for thin-layer chromatography (HPTLC/TLC). Finally, the isolated component is crystallised to get pure crystals [38].

Structural elucidation using spectroscopic methods requires perfectly pure substances—crystalline, amorphous or oily. If crystals are successfully obtained, an X-ray diffraction study could then be performed. If not, if the purified component has a non-crystalline structure, then the literature recommends obtaining ^{1}H- and ^{13}C-nuclear magnetic resonance spectra and mass spectra. Once we have the spectra and these are studied, the researcher proposes the structure of the component. This process requires a meticulous revision of the structures previously described in the literature.

Because of the immense number of known products, it is much easier to resort to a powerful database. In these days, many databases on the subject are available to the scientific community online, from the more conventional *Chemical Abstracts Registry*, now part of *Scifinder*, to more specific ones like *MarinLit®* published by the University of Canterbury (New Zealand), which encompasses all the literature published on natural marine products. Mention must also be made of the antibase (*Chemical Concepts*), which deals solely with natural products isolated from micro-organisms and higher fungi.

Once the bibliographic background has been checked and the conclusion has been drawn that the component isolated is new, a more refined structural elucidation has to be carried out. A second high-resolution mass spectrum (HRMS) is required for this to determine the exact molecular formula of both the molecule and the fragments of it that form in the apparatus' injection block. With this information, the two-dimensional structure of the new metabolite isolated can be deduced [39].

Obtaining the three-dimensional structure of molecules requires high-resolution nuclear magnetic resonance techniques. Such spectra provide data on coupling between nuclei that are close together in space and their dihedral angles, using the coupling constants J_{HH}. Complex two-dimensional resonance experiments called COSY, TOCSY, NOESY, HSQC, HMBC, DEPT-90, DEPT-135, etc. provide all the other necessary information [40, 41].

Apart from the techniques indicated above, analytical methods provide important tools in the qualitative and quantitative analysis of substances allowing us to establish their identity and the precise quantity of each component of a given mixture [36]. Instrumental techniques include high-resolution liquid chromatography (HPLC or UHPLC) and gas chromatography

(GC). Once connected to modern mass spectrometry, these tools resolve countless analytical problems (UHPLC-MS/MS or GC-MS) [21].

9. Results and discussion

A new *Paecilomyces variotii* strain was isolated from the marine habitat. The fungal biomass necessary for the chemical study was successfully produced on a laboratory scale. Twenty-eight structural groups were identified from volatile compounds, a large part of which are lipid compounds involved in the fatty acid pathway, fragments from its catabolism, terpenoids and a metabolite from the shikimic acid pathway. Two other non-volatile compounds, olein and ergosterol peroxide, were also isolated and identified using spectroscopy [42].

The screening of the biomass of a new marine-derived strain of *Penicillium roqueforti*, produced by liquid-state fermentation, led to the identification of several volatile organic compounds active in the fatty acid pathway, together with fragments produced as a result of their catabolism, terpenoids, and metabolites from the shikimic acid route. In addition, three non-volatile organic compounds: 9(11)-dehydroergosterol peroxide, 4-hydroxy-benzaldehyde and D-mannitol were isolated and identified using spectroscopy. The results have shown that this fungal strain produces no mycotoxin in the culture conditions applied and thus is useful for industrial applications where high value-added biomolecules are generated [21].

A GC-MS chemical screening on the biomass of a marine protist of the *Schizochytrium* genus enables the authors [43] to identify 24 kinds of organic compounds belonging to the *n*-alkanes, 1-alkenes, 1-alkanols, free fatty acids, methyl and ethyl esters of saturated and unsaturated fatty acids, saturated and unsaturated glycerides, wax esters, sterols, mono-, sesqui- and tri-terpenes. Thus, this organism from the base of the food chain, which accumulates so many nutrients and does not produce toxins, has been proposed as a very interesting specimen for modern functional nutrition.

The chemical constituents of the fermentation broth of the marine-derived fungus *Penicillium roqueforti* were determined. Several volatile organic compounds involved in the fatty acid pathway were identified, along with a terpene and a cyclic dipeptide. Three kinds of non-volatile metabolites were also identified by spectroscopy: alkanes, fatty acids and 1-alkanols. The results showed that the fermented broth of this fungal strain does not produce mycotoxins in the growing conditions used, which is an important factor, given the importance of this species for nutraceuticals [36].

A recent proposal is to study *Spirulina* metabolites (a blue-green alga) as a laboratory practise for bio-organic/bio-chemistry students. They propose that students tackle the separation and analysis of metabolites of nutritional interest using simple thin-layer chromatography (TLC) plates [44].

10. Conclusions and future direction

There are an estimated 22,000 known marine metabolites. Their value as potential drugs for industry—cosmetic, nutraceutical and pharmaceutical—is well documented; in fact in recent years, companies have appeared such as the Spanish company *Pharmamar*, that are trying to sustainably exploit the issue.

However, only a few marine metabolites have been developed commercially. This is perhaps due to the fact that marine environmental metabolomics is scarcely 60 years old, apart from the fact that the major bio-technology and/or pharmaceutical companies have invested very few resources in this field. But irrespective of whether or not marine metabolites have an industrial application, an understanding of their three-dimensional chemical structures and the bio-genetic pathways that living creatures use to produce them is already of great value in the field of marine chemical ecology.

Marine organisms use chemistry for many different purposes. The obvious objectives are to form cellular structures, genetic expression (DNA) and primary metabolism, which guarantees their basic welfare. There is also a secondary metabolism, controlled by enzymes, which is used by organisms to produce, accumulate and disseminate active biological substances into the environment that are essential for the survival of both the organism itself and others of the same or a different species.

For some time, these metabolites were classed under the definition of marine natural products (MNPs), but this definition is defective as it ignores the ecological function or role that they have. That is why more precise words such as allomone, kairomone or pheromone are increasingly applied to them, as we have explained in this chapter.

However, the future is promising, as there is an increasing awareness of the need to study the marine environment in-depth. The proof of this is the creation of four faculties of marine sciences in Spain in recent decades, which means that there is now a bachelor's degree in marine sciences, along with a range of Masters and PhD courses that focus on the sea as their field of study. Subjects like "Chemistry of Marine Natural Products" have suddenly appeared on our syllabuses. At the same time, our students are presenting their degree/master projects or their doctoral theses on marine environmental metabolomics, all of which augers a promising future for this exciting field of science.

Author details

Francisco Javier Toledo Marante

Address all correspondence to: franciscojavier.toledo@ulpgc.es

Chemistry Department, University of Las Palmas de Gran Canaria, Gran Canaria, Spain

References

[1] Bundy JG, Davey MP, Viant MR. Environmental metabolomics: a critical review and future perspectives. Metabolomics. 2009;5:3–21. doi:10.1007/s11306-008-0152-0

[2] McClintock JB, Baker BJ, editors. Marine chemical ecology. CRC Press, Washington DC; 2001. 624 p.

[3] Barbier M. Introduction to chemical ecology. Longman Group, United Kingdom; 1976. 144 p.

[4] Harborne JB. Introduction to ecological biochemistry. 4th ed. Academic Press; University of Reading, United Kingdom. 1993. 384 p.

[5] Reigosa MJ, Pedrol N, González L, editors. Allelopathy: a physiological process with ecological implications. Springer, Dordrecht; 2006. 656 p.

[6] Grasswitz TR, Jones GR. Chemical ecology. Encyclopedia of life sciences. John Wiley & Sons, USA; 2002.

[7] Karlson P, Lüscher M. Pheromones: a new term for a class of biologically active substances. Nature. 1959;183:55–56. doi:10.1038/183055a0

[8] Aubert M, Aubert J, Gauthier M. Telemediators in the marine environment. Marine Pollution Bulletin. 1978;9:93–95. doi:10.1016/0025-326X(78)90477-0

[9] Goldberg A, Barrow KD, Duffield AM. Identification of homologous fatty acid choline esters from trunkfish toxins by chemical ionization gas chromatography-mass spectrometry. Biomedical Mass Spectrometry. 1984;11:75–78. doi:10.1002/bms.1200110206

[10] Stallard MO, Faulkner DJ. Chemical constituents of the digestive gland of the sea hare *Aplysia californica*. II. Chemical transformations. Comparative Biochemistry and Physiology, Part B: Biochemistry & Molecular Biology. 1974;49:37–41. doi:10.1016/0305-0491(74)90219-3

[11] Kobayashi J, Murata O, Shigemori H, Sasaki T. Jaspisamides A-C, new cytotoxic macrolides from the Okinawan sponge Jaspis sp. Journal of Natural Products. 1993;56:787–791. doi:10.1021/np50095a021

[12] Matsunaga S, Fusetani N, Hashimoto K, Koseki K, Noma M, Noguchi H, Sankawa U. Bioactive marine metabolites. 25. Further kabiramides and halichondramides, cytotoxic macrolides embracing trisoxazole, from the *Hexabranchus* egg masses. Journal of Organic Chemistry. 1989;54:1360–1363.

[13] Matsunaga S. Trisoxazole macrolides from *Hexabranchus* nudibranchs and other marine invertebrates. In: Cimino G, Gavagnin M, Editors. Progress in molecular and subcellular biology, Issue Molluscs. Springer-Verlag, Berlin; 2006. pp. 241–260.

[14] Kubanek J, Fenical W, Pawlik JR. New antifeedant triterpene glycosides from the Caribbean sponge *Erylus formosus*. Natural Product Letters. 2001;15:275–285. doi: 10.1080/10575630108041292

[15] Paul VJ, Fenical W. Isolation of halimedatrial: chemical defense adaptation in the calcareous reef-building alga *Halimeda*. Science. 1983;221:747–749. doi:10.1126/science.221.4612.747

[16] Paul VJ, Van Alstyne KL. Activation of chemical defenses in the tropical green algae *Halimeda* spp. Journal of Experimental Marine Biology and Ecology. 1992;160:191–203. doi:10.1016/0022-0981(92)90237-5

[17] Look SA, Fenical W, Van Engen D, Clardy J. Erythrolides: unique marine diterpenoids interrelated by a naturally occurring di-π-methane rearrangement. Journal of the American Chemical Society. 1984;106:5026–5027. doi:10.1021/ja00329a077

[18] Dookran R, Maharaj D, Mootoo BS, Ramsewak R, McLean S, Reynolds WF, Tinto WF. Diterpenes from the gorgonian coral *Erythropodium caribaeorum* from the southern Caribbean. Journal of Natural Products. 1993;56:1051–1056.

[19] Banjoo D, Mootoo BS, Ramsewak RS, Sharma R, Lough AJ, McLean S, Reynolds WF. New erythrolides from the Caribbean gorgonian octocoral *Erythropodium caribaeorum*. Journal of Natural Products. 2002;65:314–318. doi:10.1021/np010333p

[20] Huang R, Wang B, Liu Y. Briarane-type diterpenoids from the gorgonian coral *Verrucella umbraculum*. Chemistry of Natural Compounds. 2012;48:516–517. doi:10.1007/s10600-012-0295-0

[21] Mioso R, Marante FJT, Bravo de Laguna IH, González JEG, Rodríguez JJS. Biomolecules produced in liquid-state fermentation by a marine-derived fungus, *Penicillium roqueforti*. Quimica Nova. 2014;37:260–267. doi:10.5935/0100-4042.20140045

[22] Moon B, Kaker BJ, McKlintock B. Purine and nucleoside metabolites from the Antarctic sponge *Isodictya erinacea*. Journal of Natural Products. 1998;61:116–118. doi:10.1021/NP970358H

[23] Amade P, Mallea M, Bouaiacha N. Isolation, structural identification and biological activity of two metabolites produced by *Penicillium olsonii* Bainier and Sartory. The Journal of Antibiotics. 1994;47:201–207. doi:10.7164/antibiotics.47.201

[24] Nechev J, Christie WW, Robaina R, de Diego FM, Ivanova A, Popov S, Stefanov K. Chemical composition of the sponge *Chondrosia reniformis* from the Canary Islands. Hydrobiologia. 2002;489:91–98.

[25] Chung HY, Yung IKS, Kim JS. Comparison of volatile compounds in dried scallops (*Chlamys farreri* and *Patinopecten yessoensis*) prepared by boiling and steaming methods. Journal of Agricultural and Food Chemistry. 2001;49:192–202. doi:10.1021/jf000692a

[26] Kozitskaya VN. Phenolic compounds of algae and the physiological role (review). Gidrobiologicheskii Zhurnal, 1984;20:54–65.

[27] Kamenarska ZG, Konaklieva SDD, Nicolova C, Kujungiev A, Stefanov KL, Popov SS. Volatile components of the freshwater algae *Spirogyra* and *Mongetia*. *Zeitschrift fuer Naturforschung, C:* Journal of Biosciences, 2000;55:495–499. doi:10.1515/znc-2000-7-801

[28] Callow ME, Callow JA. Attachment of zoospores of the fouling alga Enteromorpha in the presence of zosteric acid. Biofouling. 1998;13:87–95. doi: 10.1080/08927019809378373

[29] Xu Q, Barrios CA, Cutright T, Zhang-Newby B. Assessment of antifouling effectiveness of two natural product antifoulants by attachment study with freshwater bacteria. Environmental Science and Pollution Research International. 2005;12:278–284. doi: 10.1065/espr2005.04.244

[30] Caulier G, Flammang P, Gerbaux P, Eeckhaut I. When a repellent becomes an attractant: harmful saponins are kairomones attracting the symbiotic Harlequin crab. Scientific Reports. 2013;3:Article number 2639. doi:10.1038/srep02639

[31] Carte B, Faulkner DJ. Role of secondary metabolites in feeding associations between a predatory nudibranch, two grazing nudibranchs, and a bryozoan. Journal of Chemical Ecology, 1986;12:795–804.

[32] Jen J, Sehn JK, Catton K, Kramer A, Sarnelle O. Pheromone trail following in three dimensions by the freshwater copepod *Hesperodiaptomus shoshone*. Journal of Plankton Research. 2011;33:907–916.

[33] Maier I, Wenden A, Clayton MN. The movement of *Hormosira banksii* (Fucales, Phaeophyta) sphermatozoids in response to sexual pheromone. Journal of Experimental Botany. 1992;43:1651–1657.

[34] Kajiwara T. Chemistry of function of pheromones from marine organisms. Shikoku Kokenkaiho. 1999;50:7–13.

[35] Kupchan SM, Briton RW, Ziegler MF, Siegel CW. Bruceantin, a new potent antileukemic simaroubolide from *Brucea antidysenterica*. Journal of Organic Chemistry. 1973;38:178–179. doi:10.1021/jo00941a049

[36] Mioso R, Marante FJT, Bravo de Laguna IH. Chemical constituents of the fermentation broth of the marine-derived fungus *Penicillium roqueforti*. Revista Iberoamericana de Micología. 2015;32:147–152. doi:10.1916/j.riam.2014.01.004

[37] Joustra M, Söderqvist B, Fischer L. Gel filtration in organic solvents. Journal of Chromatography A. 1967;28:21–25.

[38] Leonard J, Lygo B, Procter G. Advanced practical organic chemistry. 3rd ed.CRC Press, Boca Raton; 2013.

[39] Colegate SM, Molyneux RJ. Bioactive natural products: detection, isolation and structural determination. 2nd ed.CRC Press, Boca Raton; 2007. 624 p.

[40] Crews P, Rodríguez J, Jaspars M. Organic structure analysis (topics in organic chemistry). 2nd ed. Oxford University Press, New York; 1998. 553 p.

[41] Bienz S, Bigler L, Fox T, Hesse M, Meier H. Spectroscopic methods in organic chemistry. Georg Thieme Verlag, Stuttgart; 2016.

[42] Marante FJT, Mioso R, Barrera JB, González JEG, Rodríguez JJS, Bravo de Laguna IH. Structural characterization and metabolite profiling of the facultative marine fungus *Paecilomyces variotii*. Annals of Microbiology. 2012;62:1601–1607. doi:10.1007/s13213-011-0416-1

[43] Mioso R, Marante FJT, González JEG, Rodríguez JJS, Bravo de Laguna IH. Metabolite profiling of *Schizochytrium* sp. By GC-MS, an oleaginous microbial source of biodiesel. Brazilian Journal of Microbiology. 2014;45:403–409.

[44] Bravo de Laguna IH, Marante FJT, Freire KRL, Mioso R. Laboratory exercise: extraction of nutraceuticals from *Spirulina* (Blue-Green Alga): a bioorganic chemistry practice using thin-layer chromatography. Biochemistry and Molecular Biology Education. 2015;43:366–369.

Chapter 2

Biofunctionality of Carotenoid Metabolites: An Insight into Qualitative and Quantitative Analysis

Bangalore Prabhashankar Arathi,
Poorigali Raghavendra-Rao Sowmya,
Kariyappa Vijay, Vallikannan Baskaran and
Rangaswamy Lakshminarayana

Additional information is available at the end of the chapter

http://dx.doi.org/10.5772/66210

Abstract

Epidemiological and clinical studies have shown that dietary intake of carotenoid-rich fruits and vegetables is positively correlated with reduction in age-related eye diseases, atherosclerosis, certain cancers and chronic diseases. Carotenoids consist of unique chemical characteristics and are highly vulnerable to structural modifications, leading to the formation of various derivatives under physiological conditions. The identification of these molecules is necessary before addressing their biological functions. Carotenoid metabolomics is believed to be highly complex to fingerprint due to instability and interference with complex biological matrices. Noteworthy, progress has been made in understanding carotenoid metabolism or its biotransformation in biological samples. In this regard, the chapter highlights the concept of metabolomics and their related bio-analytical techniques pertaining to the detection of carotenoids and their derived products to elucidate their bio-transformation on targeted biological functions. Further, this chapter highlights the various hyphenated analytical tools and their optimization.

Keywords: metabolomics, hyphenated technique, bio-functionality, metabolites

1. Introduction

Considerable research has made in understanding the potential role of carotenoids in plants and animals. Carotenoids are tetra-terpenoids, a group of natural compounds, found predominantly in photosynthesizing organisms such as green plants, algae and some bacteria [1]. There are over 750 carotenoids identified in nature; among them, major carotenoids like α-carotene, β-carotene, lycopene, lutein, zeaxanthin, astaxanthin, fucoxanthin and β-cryptoxanthin are well explored. In plants, carotenoids and their derivatives (apo-carotenoids) are believed to involve in photosynthesis, photo-protection, the precursor for hormones and production of aroma and flavor [2]. The diversity of carotenoids in nature is formed from the common biosynthetic pathway. The exploration of each distinctive carotenoid pigment from various dietary sources or non-dietary sources is the milestone in carotenoids biochemistry and metabolism. In animals and human, dietary ingestion is the only source of carotenoids obtained from plants (vegetables and fruits) or animals (meat, fish and poultry products) or other nutraceuticals sources. They are categorized based on biological functions as provitamin A or non-provitamin A carotenoids. Further, they classified as carotenes (β-carotene and lycopene) and xanthophylls with epoxy- (violaxanthin, neoxanthin, fucoxanthin), hydroxy- (lutein and zeaxanthin), keto- (astaxanthin and canthaxanthin) and methoxy- (spirilloxanthin) functional groups. In plants and certain microorganisms, under various environmental conditions, the biotransformation of carotenoids occurs as a necessary adaptation. Carotenoids research is an active area of research due to the characteristics of different chemical nature with unique biological function [3]. Epidemiological and clinical trials have correlated that consumption of carotenoids rich fruits and green vegetables decreases vitamin A deficiency problems, cancer, cardiovascular diseases and age-related macular degeneration [4]. Although diversified carotenoids with unique characteristics structures exist in nature, the relation between structure function needs to explore by developing appropriate analytical protocols and techniques. Only a few carotenoids are studied and detailed from past three decades, due to their presence in common dietary sources and human serum and tissues. Many of the bioactive compounds execute their function either by intact or by its derivatives [5, 6]. In many cases, carotenoids daughter molecules exist in nature as a secondary compound, or they possibly formed over series of reaction that occurs in physiological condition [7]. Carotenoids consist of long polyene chain with series of conjugated double bonds and functional groups that make them highly reactive. Thus, carotenoids are more prone to various modifications such as hydrogenation, dehydrogenation, double-bond migration, chain shortening or extension, rearrangement, isomerization, oxidation or combinations of these processes under different conditions. Apart from these, carotenoid metabolites may also form due to the presence of enzymes monooxygenase, cycloxygenase and dioxygenase [8]. Investigation on β-carotene metabolism and their conversion to vital active molecules, that is, retinyl palmitate, retinal, retinoic acids and apo-carotenoids, has made a greater contribution to the concept of metabolomics. Further, these molecules augmented with cellular function on vision, growth, immune response and related chronic health problems. With this discovery, subsequently many notable researchers worked on other non-provitamin A carotenoids and succeeded partly in the identification of bioactive metabolites in vivo. Therefore, metabolomics has attracted and motivated in an identification of carotenoid bioactive metabolites. In continuation, several studies revealed the potential role of carotenoid metabolite or oxidative products in vitro and in vivo and supported the concept that biological functions mediated through their metabolites [6–9]. With this background, we highlight the possible biotransformation of carotenoids (β-carotene, lycopene, lutein, astaxanthin, fucoxanthin and other carotenoids) and the related analytical techniques used for their determination. Also, biofuctionality of carotenoids metabolites discussed. Further, this chapter details the improvement of analytical techniques and their hyphenation for evaluation of carotenoids metabolite (beneficial or harmful effect). The advancement in analytical tools and the discovery of unknown carotenoids metabolites broaden the scope of carotenoids research. Hence, omics instruments and their methods perform untargeted and targeted profiling of carotenoids in foods and human samples. Apart from these, this chapter also emphasizes on the suitability and optimization of analytical methods to determine carotenoid metabolites.

2. Biotransformation of carotenoids

Even though more than 750 carotenoids are identified and predicted in the natural source, only a few carotenoids (β-carotene, lycopene, lutein, astaxanthin, fucoxanthin, canthaxanthin) addressed by using routine analytical techniques. Presently, carotenoids research is mainly focused on the analysis of dietary carotenoids and linked to biological functions. Since carotenoids are unstable molecules, it undergoes various modifications (discussed elsewhere), and hence, it is challenging to analyze such carotenoids or its metabolites. β-Carotene metabolites such as retinol, retinal and retinoic acid are studied extensively due to its significance in human health. Therefore, creating an awareness to study the metabolism of carotenoids and their bio-functions is currently warranted. The carotenoid metabolites may be involved in differential gene expression, cell-to-cell communication and cell differentiation. Furthermore, it is interesting to address the molecular interaction of carotenoids metabolites with free radicals, protein and lipids at cellular levels. Studies also show that carotenoids oxidative products may be involved in oxidative stress and act as pro-oxidant [10]. The identification of oxidative metabolites is considered to be more important before addressing their biological activity. Hence, rapid and sensitive analytical techniques are given priority. The bioconversion of major carotenoids such as β-carotene, lycopene, lutein, astaxanthin and fucoxanthin is discussed in this section.

2.1. Hydrocarbon carotenoids

β-Carotene is the most abundant provitamin A carotenoid found in human diet and tissues. It exerts a beneficial function in mammals, including humans, due to its ability to convert to vitamin A. Even though β-carotene considered as a safer form of vitamin A, under circumstances detrimental effects also ascribed [11]. A better understanding metabolism of the β-carotene and their derivatives of (retinoids) are still needed to unequivocally discriminate the beneficial or detrimental effects under various physiological conditions and thus enable the formulation of adequate dietary recommendations for different age groups of individuals. Symmetric oxidative cleavage of β-carotene by the enzyme β-carotene-15, 15′-monooxygenase (BCMO1) generates two molecules of retinaldehyde, and these molecules further oxidized to form retinoic acids by aldehyde dehydrogenase. The oxidation of retinoic acid conversion into more polar compound 4-oxo retinoic acid by cytochrome P450 enzymes is believed to be transcriptionally inactive. Alternatively, different forms of alcohol dehydrogenase and a variety of retinol dehydrogenases reduce retinaldehyde to retinol. Subsequently, these molecules further get esterified into retinyl esters by the enzyme lecithin retinol acyltransferase. Also, apo-carotenals can be generated by eccentric cleavage of β-carotene. The cleavage at the 9′, 10′ double bond is catalyzed by β-carotene 9′, 10′-oxygenase 2 (CMO II) and leads to the formation of β-apo-10′-carotenal and β-ionone. Apo-carotenals are ultimately converted into one molecule of retinaldehyde, and the mechanism of this conversion is not completely elucidated [12].

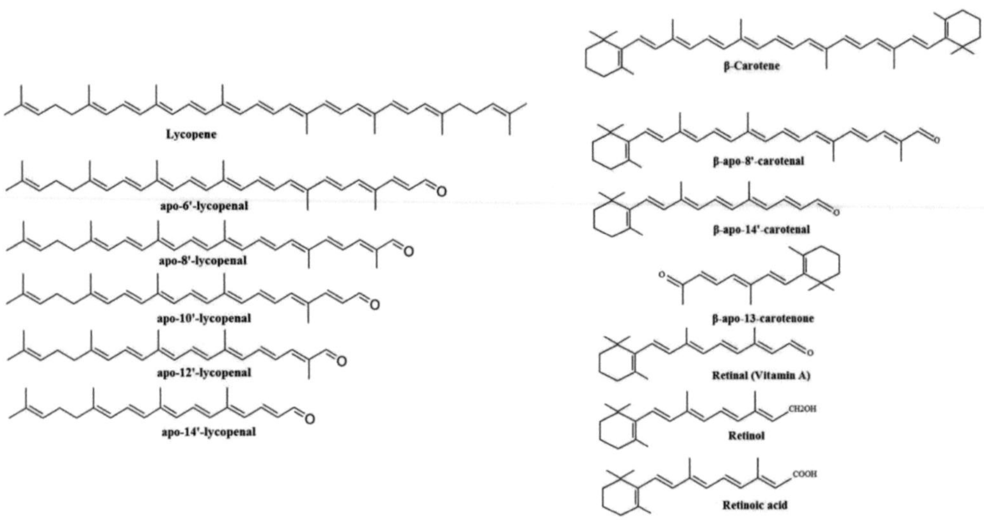

Figure 1. Molecular structures of hydrocarbon carotenoids and their metabolites.

Lycopene is another major hydrocarbon carotenoid extensively studied due to its potential role in the reduction in certain chronic diseases including cancer and cardiovascular disease. Recently, metabolism of lycopene has made a greater insight into the biological role of its derivates. These observations raised an important question about the effect of lycopene on various cellular functions and signaling pathways are a result of the direct actions of intact lycopene or its derivatives. Considerable efforts have been expended to identify its biological and physiochemical properties. β-Carotene and lycopene have the same molecular mass and chemical formula, yet lycopene is an open-polyene chain lacking the β-ionone ring structure. The metabolism of β-carotene studied extensively, but biological activities of lycopene are not detailed. Derivates of lycopene formed due to shortening the carbon chain by removal of fragments from one or both the ends. Recent studies have shown that BCMO II enzyme catalyzes lycopene eccentrically to form apo-10-carotenoids. Cleavage of cis-lycopene by BCO II may occur at either 9, 10 or 9′, 10′ double bond to produce apo-10′-lycopenal, which can be oxidized to apo-10′-lycopenoic acid or reduced to apo-10′-lycopenol. Lycopene metabolites, including apo-6-, apo-8′-, apo-10′-, apo-12′- and apo-14′-lycopenal, were detected in the plasma of humans [13]. However, the identification of cleaved lycopene metabolites in vivo is challenging. Previously, Khachik et al. [9] identified a group of lycopene oxidative products, 2,6-cyclolycopene-1,5-diol in human serum and tissues. In animal studies, several metabolites were detected in lung tissue of lycopene-supplemented ferrets. The intermediate primary cleavage product apo-10′-lycopenal could be either reduced apo-10-lycopenol or oxidized to apo-10′-lycopenoic acid. The apo-10′-lycopenal was further converted into apo-10′-lycopenoic acid in the presence of NAD+. Similarly in presence of NADH lycopene was converted into both apo-10′-lycopenoic acid and apo-10′-lycopenol [14]. Gajic et al. [15] reported apo-8′- and apo-12′-lycopenal as well as other unidentified polar metabolites of lycopene in the liver of rats when supplemented with lycopene-rich diet. Interestingly, a recent study indicated that apo-10′- and apo-14′-lycopenoic acid have a remarkable ability to upregulate BCO II expression [16]. These results showed that lycopene converted to apo-10′-lycopenoids in mammalian tissues both in vitro and in vivo. These observations presumed that lycopene metabolites play an important biological functions related to human health. The molecular structures of hydrocarbon carotenoids and their metabolites are shown in **Figure 1**.

2.2. Hydroxyl carotenoids

Lutein is one of the major xanthophylls present in green leafy vegetables. Lutein and its isomer zeaxanthin selectively accumulated in the macula of the human retina. They are recognized as antioxidants and as blue light filters [17] to protect the eyes from sunlight exposure and other lifestyle-related oxidative stress, which can lead to age-related macular degeneration and cataracts. Khachik et al. [9] identified lutein metabolites in human tissues such as plasma, milk, liver and retina. Yonekura et al. [18] observed remarkable accumulation of lutein metabolites, 3′-Hydroxy-ε,ε-caroten-3-one along with ε,ε-carotene-3,3′-dione in the plasma, liver, kidney and adipose tissues of mice fed with lutein. Further, findings indicated that mice actively convert lutein to keto-carotenoids by oxidizing the secondary hydroxyl group. 3′-oxolutein is the metabolite of lutein detected in human plasma and retina [19]. Similarly, 3′-epilutein identified in human retina and presumed that this product might form by a reduction of 3′-oxolutein from lutein. Other lutein derivatives like meso-zeaxanthin detected only in the retina and formed due to double-bond migration from lutein [20]. The dehydration products of lutein such as, 3-hydroxy-3′, 4′-didehydro-β,γ-carotene and 3-hydroxy-2′,3′-didehydro-β,ε-carotene may be formed non-enzymatically in stomach under acidic conditions [9]. Also, studies demonstrated and indicated the cleavage reaction of xanthophylls occurred in mammals by BCO II by cleaving double bond at C-9′ and C-10′ of xanthophylls [21].

2.3. Epoxy- and keto-carotenoids

The fucoxanthin is a marine carotenoid found in brown seaweeds, macroalgae and diatoms, with biological properties. The bioconversion of fucoxanthin to fucoxanthinol and amarouciaxanthin A was found in the plasma and liver of mice fed with fucoxanthin, whereas fucoxanthin was not detected [22]. Fucoxanthinol hydrolyzed from fucoxanthin in the intestinal tract, circulated in the body and then oxidatively converted into amarouciaxanthin A. The conversion of fucoxanthinol into amarouciaxanthin A found in human hepatoma HepG2 cells. Moreover, oxidative conversion of xanthophylls in mouse liver microsomal fractions required NAD^+ as a co-factor, and this result demonstrates the metabolic conversion of the 3-hydroxyl end group at the level of enzyme reaction [23].

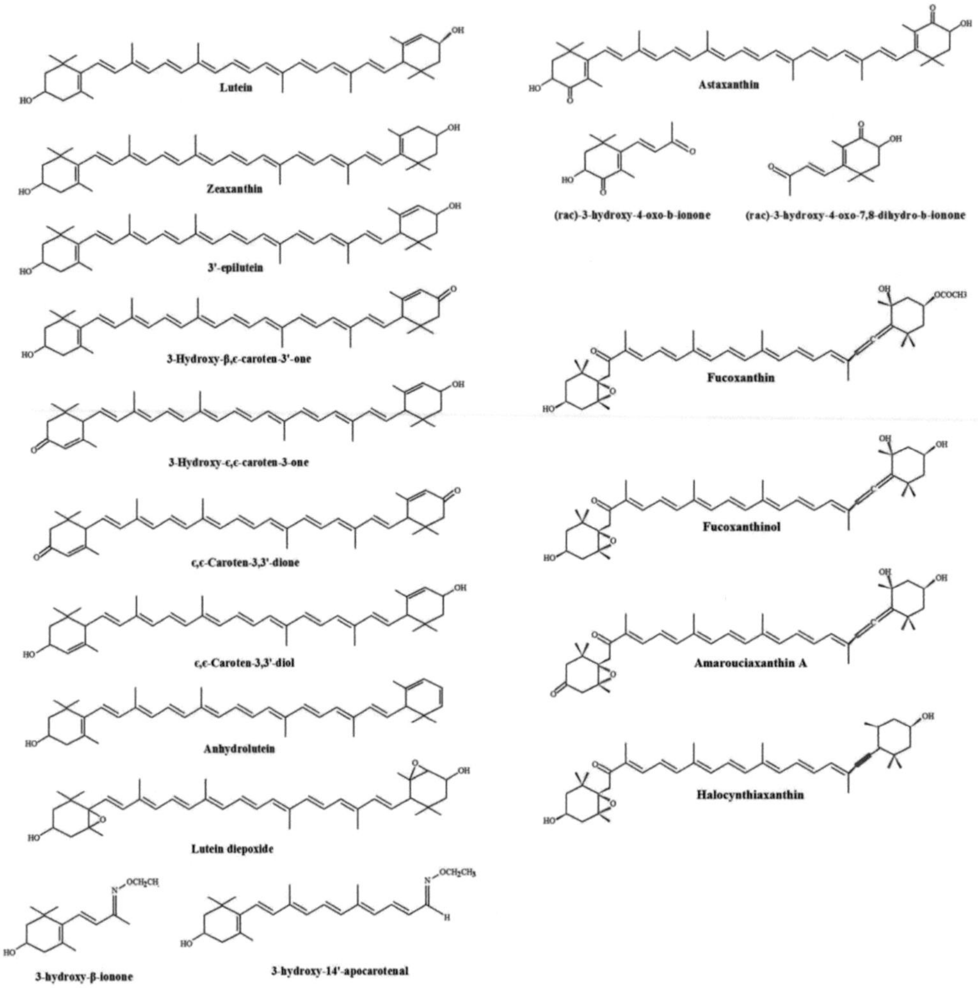

Figure 2. Molecular structures of hydroxyl-, epoxy- and keto-carotenoids and their metabolites.

Astaxanthin (ASTX) is a major keto-carotenoid metabolized into 3-hydroxy-4-oxo-β-ionone and 3-hydroxy-4-oxo-7,8-dihydro-β-ionone in primary rat hepatocytes [24]. However, enzymes involved in the synthesis of these metabolites and their potential biological functions are not elucidated. However, ASTX incubated with microsomes containing cytochrome P450 (CYPs) did not generate ASTX metabolites and induction of CYP activity in hepatocytes [25]. Therefore, the CYPs enzymes are not responsible for the production of ASTX metabolites. Hence, investigation on ASTX metabolites and their metabolism needs to explore. However, there are no detailed in vivo studies available on astaxanthin metabolites except isomers detected in human serum samples [26]. Canthaxanthin, (4,4′-diketo-β,β-carotene), is a keto-carotenoid found in green algae, crustaceans and certain microorganism. This pigment also used as poultry feed to extend the dominant pigments in the skin and egg yolk of chickens. Analysis of tissues from chicks revealed that a portion of canthaxanthin reduced to 4-hydroxyechinenone (4-hydroxy-4′-keto-β,β-carotene) and iso-zeaxanthin (4,4′-dihydroxy-β,β-carotene). Ratio of canthaxanthin to metabolites depends on type of tissue, but in general metabolites concentrated in the skin [27]. The molecular structure of hydroxyl-, epoxy- and keto-carotenoids and their metabolites is shown in **Figure 2**.

3. Sample preparation for carotenoids and their metabolites

Biological samples used for analysis of carotenoid and their metabolites need to store in −80°C under argon or nitrogen gas to prevent carotenoids deterioration. Before analysis, the tissues samples thawed at room temperature before homogenization using suitable solvents, and then, organic phase collected after vortexing and centrifugation. To this appropriate amount of anhydrous, sodium sulfate should be added to remove traces of moisture. The organic phase evaporated with argon or nitrogen gas, and the extract was reconstituted in mobile phase and injected into the LC/MS/MS equipment. Other than the sample preparation adequate care should be taken to avoid light, heat and exposure to atmospheric oxygen, and the high quality solvents used for extraction and analysis, proper storage of samples under inert conditions, addition of antioxidants and completion of analysis within short run time applied with sophisticated instruments.

4. Analysis of carotenoids and their isomers, cleavage products/oxidation products or metabolites in food and biological samples

HPLC with C18 and C30 stationary phases either with reverse phase or normal phase used extensively for the analysis of a diverse group of carotenoids in various natural sources including food and biological samples [28, 29]. In general, the separation and resolution of carotenoids are better with gradient than isocratic solvents system. However, these methods require higher analysis time and solvent consumption. Further, these conditions and stationary phase were successfully employed in the analysis of carotenoids and their isomers or/and related derived products in biological fluids and tissues samples [30, 31]. Others have attempted and identified several types of carotenoids and their metabolites/oxidative products by employing gas chromatography and mass spectrometry. Stratton et al. [32] determined β-carotene oxidation products as β-ionone, β-apo-14′-carotenal, β-apo-10′-carotenal, β-apo-8′-carotenal and β-carotene 5,8-endoperoxide by using RP-HPLC and GC-MS. Wyss and Bucheli [33] developed an HPLC method with automated column switching for the simultaneous determination of endogenous levels of 13-cis-retinoic acid, all-trans-retinoic acid and their 4-oxo metabolites in human and animals tissue samples. Khachik et al. [9] separated, identified, quantified and compared 34 carotenoids, including 13 geometrical isomers and 8 metabolites in breast milk and serum of lactating mothers by HPLC-

PDA-MS. Wolz et al. [24] investigated the astaxanthin metabolites in primary cultures of rat hepatocytes by GC-MS. Siems et al. [34] demonstrated the oxidation products of β-carotene by using capillary gas-liquid chromatography and HPLC. The method developed was linear in the range of 0.3–100 ng/mL with a lower quantification limit. Kim et al. [35] isolated and analyzed auto-oxidation products of lycopene by GC-MS. Subsequently, Bernstein et al. [36] identified and quantified the dietary lutein and their geometrical (E/Z) isomers and related metabolites by HPLC in tissues of the human eye. Dachtler et al. [37] identified carotenoid stereoisomers in spinach and human retina samples by using HPLC online coupled to mass spectrometry and nuclear magnetic resonance spectroscopy. Aust et al. [38] have reported lycopene oxidative product 2,7,11-trimethyltetradecahexaene-1,14-dial by using GC-MS. Sommerburg et al. [39] separated and identified 5,6-epoxi-β-ionone, ionene, β-cyclocitral, β-ionone, dihydroactinidiolide and 4-oxo-β-ionone as major cleavage products of β-carotene mediated by hypochlorous acid using GC-MS.

Simultaneously from the late 2000 s, spectroscopic and mass spectrometric techniques have being used for qualitative and quantitative analysis for structurally different carotenoids and their derived products. The carotenoid analysis is done by using different ionization modes such as matrix-assisted laser desorption/ionization (MALDI), electrospray (ESI), atmospheric pressure chemical ionization (APCI) and atmospheric pressure photoionization (APPI). These methods are employed successfully to target the mass with better ionization and high resolution for the identification of broad range of natural organic molecules including carotenoids [40–42]. Further, this application is extended to explore several metabolites and oxidative products of carotenoids in rodents and human biological samples [18, 43–45]. Initially, van Breemen [46] developed an APCI-LC-MS method and analyzed all-trans-retinol and all-trans-retinyl palmitate with the lower limit of detection (LOD) of 34 fmoles/μL and 36 fmoles/μL (on-column), and limit of quantitation (LOQ) was 500 fmoles/μl and 250 fmoles/μl (on-column), respectively. Later, Zhu et al. [47] developed LC/APCI-MS a negative ion mode for the measurement of labeled and unlabeled β-carotene in human serum and feces to demonstrate bioefficacy of orally administered β-carotene, and the limit of detection of 0.25 pmol (on injection of 20 μL of 0.0125 μM β-carotene) and LOQ was 1.0 pmol (on the injection of 20 μL of 0.050 μM β-carotene) with the linear range of 1.1–2179 pmoles on-column. Further, they believed that linear range with low LOD and LOQ facilitated sensitive and selective analysis of provitamin A carotenoids. Fraser et al. [40] used MALDI/TOF-MS to detect and quantify plant carotenoids, and other metabolites (m/z) in complex biological systems. Schäffer et al. [48] validated RP-HPLC-DAD method for simultaneous quantification of retinol, retinyl esters, tocopherols and selected carotenoids in the lung, liver and plasma of various animal samples. Würtinger and Oberacher [49] demonstrated influence of analytical parameters like constituents of mobile phase, including the modifiers added (acids, bases, dopants, metals and salts) and other experimental conditions (collision energy, flow collision gas, temperatures, etc.) on the ionization of the analyses (e.g., ESI, APCI, APPI, FAB). Consequently, Giuffrida et al. [50] for the first time analyzed and identified 52 carotenoids among the various cultivars of Capsicum using HPLC-DAD-APCI-MS. Further, Rivera et al. [51] studied the effect of ionization of carotenes, oxygenated carotenoids using ESI, APCI and APPI and reported that 12 of the 16 carotenoids exhibited strongest signal strength with APCI.

Furthermore, MS/MS is considered to be an efficient method for carotenoids identification by the use of transitions through precursor and daughter ions. This approach is also suitable for the determination of carotenoids with the same molecular mass such as structural isomers. Prasain et al. [52] developed a sensitive and specific ESI-MS/MS method using MRM for the detection of carotenoid oximes (zeaxanthin oxidation products) in a human eye sample. Further, this provides invaluable information on the characterization and quantification of carotenoids and their oxidized products formed during in vitro and in vivo studies. Crupi et al. [41] investigated carotenoids in a typical wine grape variety by using RP-HPLC-DAD-MS (ESI^+) method. Due to an unusual ionization process, their mass spectra of carotenoids comprised both protonated molecules and molecular ion radicals. Further, they were subjected for the selective collision-induced dissociation (CID) to differentiate structural and geometrical isomers such as lutein isomers (zeaxanthin, 9Z and 9′Z-lutein) and a cis-isomer of β-carotene (9Z- β-carotene), 5,6-epoxy xanthophylls, violaxanthin, (9′Z)-neoxanthin, lutein-5,6-epoxide and 5,8-epoxy xanthophylls diastereoisomers (neochrome, auroxanthin, luteoxanthin, flavoxanthin, chrysanthemaxanthin). Kopec et al. [13] performed HPLC-MS/MS using $APCI^{-ve}$ mode to separate and detect the apo-6′-, apo-8′-, apo-10′-, apo-12′-, apo-14′- and apo-15′-lycopenal products formed by in vitro oxidation reaction. The quantitative analysis of carotenoids and other fat-soluble compounds performed simultaneously on a C30 column and detected by APCI-MS/MS with operation of selected reaction monitoring (SRM) mode. Also, this method calibrated to shown a less variability in intra- and inter-day precision analysis. Recently, ultrahigh performance liquid chromatography (UPLC/UHPLC) has attracted much attention due to faster analysis and higher sensitivity with 2 μm particle size stationary phases, thereby increasing column efficiency, decreasing band broadening, and increasing resolution [4]. Granado-Lorencio et al. [53] assessed the suitability of UHPLC for the simultaneous determination of biomarkers of vitamins including vitamin A (retinol, retinyl esters) and major carotenoids in human serum. This method allowed a better resolution for carotenes and xanthophylls isomers provides better sensitivity and reproducibility in peak area and retention time than the HPLC, with mean RSDs. Rivera et al. [51] analyzed various carotenoids by UHPLC-MS/MS detection. Further, they compared three different ionization techniques (ESI, APCI and APPI) to ionize the carotenoids and concluded that APCI has a powerful technique to ionize carotenoids. They also used dopants (acetone, toluene, anisole and chlorobenzene) that allowed the enhancement of the carotenoid signals strength up to 178-fold. Delpino-Rius et al. [54] analyzed simultaneously epoxy carotenoids, hydroxyl carotenoids and carotenes in fresh homemade and industrially processed fruit products by UPLC. They identified 27 carotenoids eluted within 17 min; furthermore, this method allowed to differentiate the carotenoid profiles and 5,6- to 5,8-epoxycarotenoids. Separation of carotenoids on UHPLC columns illustrates less analysis time compared to HPLC C30 column; however, the separation is better in C30 column. Hence, there is a requirement for appropriate UHPLC column for rapid and sensitive analysis of carotenoids with better separation. Zhao et al. [55] developed a quick and simple ultrahigh performance supercritical fluid chromatography-photodiode array detector (UHPSFC-PDA) method and validated the determination of carotenoids in dietary supplements using Acquity UPC2HSS column by gradient elution with carbon dioxide and solvent system. We have developed a rapid UPLC-MS/MS method for analysis of lycopene isomers and their fragmentation pattern with CID and demonstrated the importance of ion mobility to differentiate carotenoids geometrical isomers [56]. Dong et al. [57] characterized and distinguished geometrical isomers of lycopene and β-carotene using ion mobility spectrometry. Raphael et al. [58] identified canthaxanthin oxidation products and compared the similarity of β-carotene like oxidation products by using both LC–MS and GC–MS chromatograms.

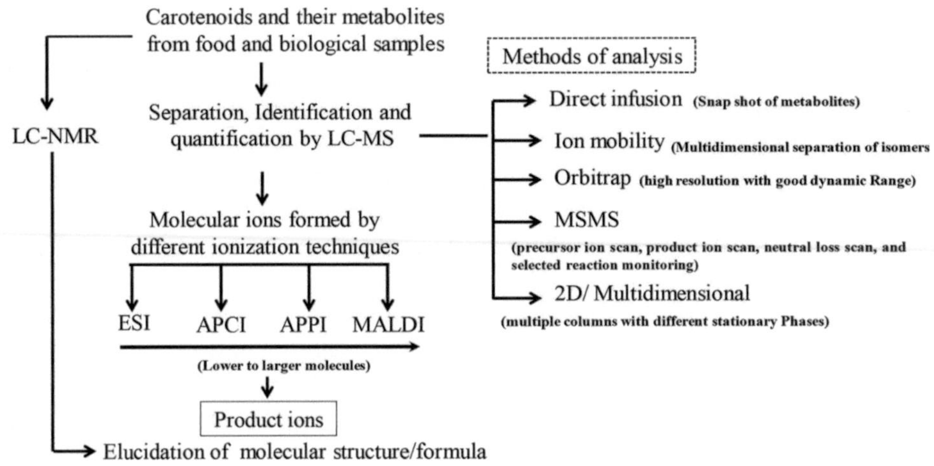

Figure 3. Hyphenated analytical techniques for the characterization of carotenoids and their metabolites. The details of these techniques are discussed in Section 4.

Apart from these, Orbitrap MS a high-resolution mass spectrometry was also exploited for carotenoids analysis to generate mass spectra with a resolving power up to 100,000 at full-width half-maximum and mass accuracies within two parts per million (ppm). Due to its high mass resolution and exact mass screening detectors, probable molecular formulae of the ions and fragments were elucidated [59]. In continuation, Van Meulebroek et al. [60] developed a full-scan high-resolution Orbitrap MS method enabling the metabolomic screening for carotenoids in tomato fruit tissue. The validation demonstrated the excellent performance in terms of linearity, repeatability and higher range of mean corrected recovery. Additionally, a well-established detection technique, that is, MS/MS and ultraviolet-visible spectroscopy photodiode array, indicated superior performance of high-resolution Orbitrap MS (with limits of detection ranging from 1.0 to 3.8 pg μL^{-1}). Contemporarily, 2D-LC and multi-dimensional chromatography have emerged as a tool for carotenoid analysis. This provides an excellent separation and resolution for analysis of complex matrices. In this regard, Cacciola, et al. [61] developed and applied a comprehensive normal-phase × reversed-phase liquid chromatography (NP-LC × RP-LC) system for analysis of the intact carotenoid composition of red chili peppers, with photodiode array and mass spectrometry detection. A total of 33 compounds separated into 10 diverse chemical classes in the 2-D space and identified by accurate IT-TOF (ion trap-time of flight) MS. Apart from these, the robust technique LC-NMR offers 1-D and

2-D NMR spectra for the components separated by HPLC. Recently, LC-NMR is used because of improved sensitivity due to higher magnetic fields. Further, NMR provides information about conformational geometry for structural elucidation. LC-NMR is established as a method of analyzing major carotenoids and metabolites in food and biological samples [62]. The NMR studies on carotenoids metabolites are scanty due to quantitative limitations. Therefore, LC-MS-APCI studies are widely used in the characterization of carotenoids metabolites or oxidative cleavage products. The approach of LC-MS techniques and ionization modes for characterization of carotenoids and their metabolites are shown in **Figure 3**.

Even advancement in hyphenated analytical techniques and inconsistency of results may arise due to several pre-chromatographic (samples or tissues, the nature of carotenoids, sample preparation, incomplete extraction, solvent incompatibility, isomerization/oxidation, physical losses of carotenoids/metabolites and its accountability) and post-chromatographic (low recovery, less stability, inaccurate method validation, co-elution, unavailability of standards, selection of improper mode of ionization, carotenoid/metabolites with same molecular mass) errors in the carotenoids and their metabolites analysis [4]. The qualitative and quantitative analyses of carotenoids need a proper method validation as per the ICH (International Council on Harmonization of Technical Requirements for Registration of Pharmaceuticals for Human Use) and IUPAC (International Union of Pure and Applied Chemistry) guidelines. The method validation mainly comprises of precision, specificity, LOD, LOQ, linearity, sensitivity, range and robustness. Carotenoid standard curves prepared for subsequent quantitation by HPLC–PDA. The amounts of carotenoids calculated based on the regression equations. The LOD and LOQ calculated for each standard on the basis of the signal-to-noise ratio (3:1 for LOD and 10:1 for LOQ). LOD is the amount that resulted in a peak with a height three times that of the baseline noise, respectively, and the LOQ determined as lowest injected amount which could be quantifiable reproducibility (RSD $\leq$ 5%). The precision evaluated by the relative coefficient of variation (%). The inter-and intra-day relative standard deviations (RSD) for retention times of individual carotenoid are considered for standard concentration to check the reproducibility of the method. The accuracy of the extraction method is assessed by determining recovery of carotenoids with a mean value, respectively.

5. Biofunctionality of hydrocarbon carotenoid metabolites

β-Carotene is one of the most potent vitamin A precursor among other provitamin A carotenoids (α-carotene, β-cryptoxanthin and γ-carotene). Chemically, vitamin A refers to all isoprenoid compounds that possess the biological activity of all-trans-retinol. The parent structure of most retinoids contains a substituted β-ionone ring with a side chain of three isoprenoid units linked at the 6-position of the β-ionone ring. Retinol plays a role in vision, differentiation and proliferation of a wide range of epithelial cells, bone growth, reproduction and embryonic development. β-Carotene conversion to retinal, retinoic acid and other active forms of the vitamin A family is well documented. Retinoic acids serve as ligands for nuclear retinoic acid receptors (RARs), namely RXR and RAR, mediate vitamin A dependent activities. Several studies have reported the biological properties of β-apocarotenoids. Zile et al. [63]

demonstrated that retinol β-D-glucuronide (a metabolite of vitamin A) could arrest HL-60 cell proliferation and induces differentiation into mature granulocytes; it may act by itself or by being hydrolyzed to retinoic acid. Others have reported that β-carotene cleavage products, such as β-apo-14′-carotenoic acid, stimulate the differentiation of U937 leukemia cells, and β-apo-12′-carotenoic acid inhibits the proliferation of HL-60 cells [64, 65]. Hu et al. [66] characterized a polar oxidation product of β-carotene 5,8-endoperoxy-2,3-dihydro-β-apocarotene-13-one and demonstrated the inhibition of cell growth and cholesterol synthesis in MCF-7 cells. Kawada et al. [67] suggested that the inhibitory action of adipocyte differentiation by carotenoids and retinoids exhibited through the RAR upregulation and the suppression of PPARγ. Tibaduiza et al. [68] tested the effect of synthetic eccentric cleavage products of β-carotene such as β-apocarotenoid acids, including β-apo-8′-, β-apo-10′-, β-apo-12′- and β-apo-14′-carotenoic acid on the growth of the human breast cancer cells. Further, they observed β-apo-14′ and β-apo-12′-carotenoic acid significantly inhibited MCF-7 growth, and only β-apo-14′-carotenoic acid inhibited Hs578T growth. Further, they observed none of these treatments inhibited the growth of MDA-MB-231 cells. Also, authors concluded that apocarotenoid acids exhibit antiproliferative effects by downregulation of cell cycle regulatory proteins and inhibition of AP-1 transcriptional activity. Rühl et al. [69] also reported that β-carotene is involved in activation of human pregnane receptor (PXR) a ligand-activated transcription factors involved in xenobiotic detoxification in the liver. Ziouzenkova et al. [70] demonstrated that β-apo-14′-carotenal, but not other structurally related apo-carotenals, represses peroxisome proliferator-activated receptors (PPAR) and RXR activation. Eroglu et al. [71] demonstrated that none of the β-apocarotenoids significantly activated RARs. However, β-apo-14′-carotenal, β-apo-14′-carotenoic acid and β-apo-13-carotenone antagonized ATRA-induced transactivation of RARs.

Lycopene is another non-provitamin A hydrocarbon carotenoid present in human serum and tissues. Recently, lycopene has attracted much attention due to its association with a decreased risk of certain cancers, cardiovascular diseases and other chronic problems [72]. Several studies from in vivo and in vitro suggest that lycopene induces apoptosis in cancer cells [73–75]. Structurally, lycopene consists of open-polyene chain lacking the β-ionone ring compared to β-carotene and share a same molecular mass and chemical formula. Even though the metabolism of β-carotene extensively studied, the metabolism and biological activities of lycopene are not detailed. Several studies support the concept that biological activities of lycopene are mediated by its oxidative products/cleavage products [76–78]. The characterization of BCO II enzyme has demonstrated that this enzyme can catalyze the eccentric cleavage of lycopene to form apo-10′-lycopenoids [14]. Several lycopene metabolites have been identified in vivo and in vitro [9, 13, 15, 35, 79], but their biological activities and mechanism of action need elucidation. The discovery of various oxidation products or metabolites of carotenoids has questioned the active role of them compared to intact carotenoid molecules in combating various diseases. However, metabolites or oxidative products of LYC on health benefits are warranted. Nara et al. [80] have demonstrated that autoxidation mixtures of LYC inhibited the HL-60 cell growth effectively than LYC. Similarly, Zhang et al. [81] identified a cleavage product of LYC (E, E, E)-4-methyl-8-oxo-2,4,6-nonatrienal and evaluated its apoptosis-inducing activity in HL-60 cells. Consequently, Aust et al. [38] reported the role of LYC degraded products in enhancing cell communication and cell signaling. Lian et al. [5] demonstrated that apo-10′-lycopenoic acid is a biologically active metabolite of lycopene with potential chemopreventive agent against lung tumorigenesis. Ford et al. [82] demonstrated that lycopene and apo-12′-lycopenal reduce the proliferation of prostate cancer cells. In the recent study, apo-10′-lycopenoic acid led to the increase in SIRT1 (Sirtuin 1) enzyme activity by treatment with this metabolite in mice, resulting in prevention of fatty liver [83]. Due to its high antioxidant activity, lycopene is more unstable and needs more advanced analytical techniques to identify and characterize lycopene metabolites and their biological activities.

6. Biofunctionality of oxygenated carotenoid metabolites

Lutein and its isomer zeaxanthin are the two major carotenoids found in the human eye associated with vision protective properties. Dietary consumption of lutein-rich fruits and vegetables positively associated in decreasing AMD and cataract [8, 17]. Although conversions of β-carotene to retinoids documented in animals and humans, only little is known about the metabolism of xanthophyll carotenoids. In general, carotenoid metabolites are involved in chemoprevention of cancer [6]. Lutein metabolites such as 3′-epilutein, 3′-oxolutein, 3′-dehydrolutein, meso-zeaxanthin, methoxy-zeaxanthin, oxime derivatives of 3-hydroxy-β-ionone and 3-hydroxy-14-apocarotenal reported in human tissues and serum [9, 43, 52]. Characterization of these lutein and zeaxanthin metabolites in vitro and in vivo is warranted to address their biological functions. The health benefit of these metabolites is not detailed. The oxidized lutein may be highly reactive when combined with similar reactive oxygen species and presumed to enhance antioxidant property [6], but the mechanism of action of these xanthophylls (lutein) metabolites remains unanswered. Several noteworthy studies have explored lutein metabolites in vivo [9, 18, 43, 44, 52]. However, mechanism of action and other functional aspects of these metabolites/oxidative cleavage products need further research. We reported the possible protective effect of lutein oxidation products in cervical cancer cell lines [84]. Further, we also elucidated the formation of lutein oxidation products mediated through peroxyl radicals and screened the antioxidant and cytotoxic effects of oxidized lutein on HeLa cancer cells [6]. Previously, we identified the apocarotenals, diepoxides and other oxidative degradation products of lutein in liver. Further, we presumed that these products are formed ay due to the peroxyl radical-mediated oxidation in the body [6, 44]. These results are significant in chain breaking peroxyl radical or quenching of singlet oxygen. Furthermore, the existence of these oxidized molecules in vivo is important in free radical chemistry and oxidative stress [85]. Possibly, oxidized lutein may reduce the cancer cells viability through induction of apoptosis. Further, result from our study demonstrated the inhibitory effect of these compounds on cancer cell growth, which may be due to the effect on signaling pathway involved in apoptosis. The biological activity of intact lutein may be different than its metabolites. Hence, it is important to address that the beneficial role of lutein and zeaxanthin in delaying and possibly protecting against ascribed chronic diseases may be due to their metabolites. In the case of other oxygenated carotenoids, such as astaxanthin, Wolz et al. [24] identified the metabolized products such as 3-hydroxy-4-oxo-β-ionone and 3-hydroxy-4-oxo-7,8-dihydro-β-ionone in primary rat hepatocytes. Later, Kistler et al. [25] reported four radiolabeled metabolites of astaxanthin including 3-hydroxy-4-oxo-β-ionol, 3-hydroxy-4-oxo-β-ionone, 3-hydroxy-4-oxo-7,8-dihydro-β-ionol and 3-hydroxy-4-oxo-7,8-dihydro-β-ionone in human and primary human hepatocytes. Further, they demonstrated that incubation of astaxanthin with microsomes from liver containing detoxifying enzymes did not generate astaxanthin metabolites. Sangeetha and Baskaran [86] hypothesized that astaxanthin might be converted into retinol via β-carotene in retinol-deficient rats. However, the formation of astaxanthin metabolites and their biological functions is not detailed. Further, studies related to astaxanthin metabolites and their biological functions are not detailed and explored.

Fucoxanthin is a major epoxy carotenoid found in the marine source (seaweeds) and explored for its anticancer, anti-allergic and anti-obese activities [87, 88]. Dietary fucoxanthin is metabolized to fucoxanthinol in the gastrointestinal tract by digestive enzymes, and further, it is converted to amarouciaxanthin in liver [23]. The bioactivity of fucoxanthin is attributed to its metabolites fucoxanthinol and amarouciaxanthin A. There are reports demonstrated that fucoxanthinol is effective and plays an important role in health benefits than intact fucoxanthin [89, 90]. Anti-proliferative and cancer preventing influences of fucoxanthin and fucoxanthinol are mediated through different signaling pathways [90]. Also, others have reported fucoxanthin, and its metabolites regulate adipogenic gene expression and inhibit the adipocyte differentiation of 3T3-L1 cells through downregulation of PPARγ. The suppressive effect of fucoxanthinol is superior on adipocyte differentiation than its parent molecule [91]. In MDA-MB-231 cells, fucoxanthinol reduced nuclear levels of NF-κB members and indicated an effective for the treatment and/or prevention of breast cancer [89]. However, there are no much reports on epoxy carotenoids identified in blood and tissues and may be due to the less dietary importance. This is an active area of research and deserves further study. The overview of biofunctionality of carotenoid cleavage products/metabolites is shown in **Table 1**.

Carotenoid	Metabolites identified	Bio-functionality of carotenoid metabolites	References
Lycopene	Oxidation products of lycopene	Inhibits the growth of HL-60 human promyelocytic leukemia cells	[80]
	Acyclo retinoic acid	Activates RAR to inhibit mammary cancer cells growth	[92]
	2,7,11-Trimethyl-tetradecahexaene-1,14-dial	Enhance gap-junction communication	[38]
	(E,E,E)-4-methyl-8-oxo-2,4,6-nonatrienal (MON)	Induce apoptosis by downregulation of Bcl-2 and Bcl-XL, and activated caspase cascades in HL-60 cells	[81]
	Apo-10′-lycopenoic acid	Chemopreventive effect on lung carcinogenesis	[5]
	Apo-12′-lycopenal	Reduced androgen-independent prostate cancer cells	[82]
	Apo-10′-lycopenoic acid	Influence on SIRT1 enzyme activity and prevention of fatty liver	[83]

http://dx.doi.org/10.5772/66210

Carotenoid	Metabolites identified	Bio-functionality of carotenoid metabolites	References
β-Carotene	5,8-Endoperoxy-2,3-dihydro-β-apocarotene-13-one	Inhibits cell growth and cholesterol synthesis in MCF-7 cells	[66]
	5,6-Monoepoxide	Greater differentiation-inducing activity than β-carotene toward NB4 human leukemia cells	[93]
	Retinoyl β-D-glucuronide	Arrest HL-60 cell proliferation and induce their differentiation into mature granulocytes	[63]
	β-Apo-14′-carotenoic acid	Stimulates the differentiation of U937 leukemia cells	[64]
	β-Apo-12′-carotenoic acid	Inhibits the proliferation of HL-60 cells	[65]
	Retinoids	Inhibits adipocyte differentiation through RAR upregulation and suppression of PPARγ	[67]
	β-Apo-14′ and β-apo-12′-carotenoic acid	Inhibits MCF-7 growth, and Hs578T growth	[68]
	β-Apo-14′-carotenal	Represses PPAR and RXR activation	[70]
	β-Apo-14′-carotenal, β-apo-14′-carotenoic acid, and β-apo-13-carotenone	Antagonized ATRA-induced transactivation of RARs	[71]
Lutein	Photodegraded and peroxyl mediated lutein oxidized products	Higher antioxidant and cytotoxic effects in HeLa cells	[6, 84]
Fucoxanthin	Fucoxanthinol and amarouciaxanthin A	Anti-proliferative effect on human prostate cancer cells MCF-7 and MDA-MB-231 cells	[23] [89]
	Halocynthiaxanthin and fucoxanthinol	Induces apoptosis in human leukemia, breast and colon cancer cells	[94]
	Fucoxanthinol	Fucoxanthin metabolites reduced the growth of primary effusion lymphoma tumor cells	[95]
Canthaxanthin	4-Oxoretinoic acid	Induction of gap junctional communication in murine fibroblasts	[96]

Table 1. Biofunctionality of carotenoid cleavage products/metabolites.

7. Conclusion

In concern to the application of analytical techniques in carotenoid, research needs a more sophisticated instrumentation to increase sensitivity, precision, specificity and speed of analysis. The increase in analytical hyphenation considered as cutting edge with multi-dimensional or techniques that may support to decipher the carotenoids found in natural sources and their effects on human health. Further, metabolomics corresponds to prevailing analytical platforms to obtain detailed and complete information on the composition of food components and their existence in biological entities. Further, the target in carotenoid/

metabolites analysis is to understand the role of these compounds at the molecular levels (i.e.,their interaction with genes and their subsequent effect on proteins and metabolites). This information will provide a rational design of strategies to manipulate cell functions through diet/nutraceuticals, which is expected to have an extraordinary impact on human health. The development of the framework in genomics, transcriptomics, proteomics and metabolomics has given rise to opportunities for increasing our understanding of different issues that can be addressed by profiling carotenoid metabolites.

Acknowledgements

Authors acknowledge Department of Science and Technology, Govt. of India, Grant Reference numbers (WOS-A: F.NO.SR/WOS-A/LS-35/2012) and (SB/EMEQ-233/2013). Authors also acknowledge the Department of Biotechnology-Bangalore University for their encouragement and support.

Author details

Bangalore Prabhashankar Arathi[1], Poorigali Raghavendra-Rao Sowmya[1], Kariyappa Vijay[1], Vallikannan Baskaran[2] and Rangaswamy Lakshminarayana[1*]

*Address all correspondence to: rlnarn21@gmail.com

1 Department of Biotechnology, Jnana Bharathi Campus, Bangalore University, Bangalore, India

2 Department of Biochemistry and Nutrition, Central Food Technological Research Institute, CSIR, Mysuru, India

References

[1] G. Britton, S. Liaaen-Jensen, H. Pfander. Carotenoids: Nutrition and Health. Birkhauser Verlag Basel, Vol. 5 (2009).

[2] N. Nisar, L. Li, S. Lu, N.C. Khin, B.J. Pogson. Carotenoid metabolism in plants. Molecular Plant, 8 (2015), 68–82.

[3] G. Britton. Structure and properties of carotenoids in relation to function. FASEB J, 9 (1995), 1551–1558.

[4] B.P. Arathi, P.R. Sowmya, K. Vijay, V. Baskaran, R. Lakshminarayana. Metabolomics of carotenoids: The challenges and prospects – A review. Trends Food Sci Tech, 45 (2015), 105–117.

[5] F. Lian, D.E. Smith, H. Ernst, R.M. Russel, X.D. Wang. Apo-10′-lycopenic acid inhibits lung cancer cell growth in vitro, and suppresses lung tumorigenesis in A/J mouse model in vivo. Carcinogenesis, 28 (2007), 1567–1574.

[6] R. Lakshminarayana, G. Aruna, U.V. Sathisha, S.M. Dharmesh, V. Baskaran. Structural elucidation of possible lutein oxidation products mediated through peroxyl radical inducer 2,2′-azobis (2-methylpropionamidine) dihydrochloride: antioxidant and cytotoxic influence of oxidized lutein in HeLa cells. Chem Biol Inter, 203 (2013), 448–455.

[7] J.A. Olson. Absorption, transport and metabolism of carotenoids in humans. Pure App Chem, 66 (1994), 1101–1106.

[8] K.J. Yeum, Y.C. Leekim, S. Yoon, K.Y. Lee, I.S. Park, K.S. Lee, B.S. Kim, G.W. Tang, R.M. Russell, N.I. Krinsky. Similar metabolites formed from β-carotene by human gastric-mucosal homogenates, lipoxygenase, or linoleic-acid hydroperoxide. Archives of Biochemistry and Biophysics, 321 (1995), 167–174.

[9] F. Khachik, C.J. Spangler, J. Cecil Smith. Identification, quantification, and Relative concentrations of carotenoids and their metabolites in human milk and serum. Anal Chem, 69 (1997), 1873–1881.

[10] W. Siems, O. Sommerburg, L. Schild, W. Augustin, C.D. Langhans, I. Wiswedel. β-carotene cleavage products induce oxidative stress in vitro by impairing mitochondrial respiration. FASEB J, 10 (2002), 1289–91.

[11] G.S. Omenn, G.E. Goodman, M.D. Thomquist. Effects of a combination of β-carotene and vitamin A on lung cancer and cardiovascular disease. New Eng J Med, 334 (1996), 1150–1155.

[12] A. Eroglu, E.H. Harrison. Carotenoid metabolism in mammals, including man: formation, occurrence, and function of apocarotenoids. J Lipid Res, 54 (2013), 1719–1730.

[13] R.E. Kopec, K.M. Riedl, E.H. Harrison, R.W. Curley, D.P. Hruskewycz, S.K. Clinton, et al. Identification and quantification of apo-lycopenals in fruits, vegetables, and human plasma. J Agri Food Chem, 58 (2010), 3290–3296.

[14] K.Q. Hu, C. Liu, H. Ernst, N.I. Krinsky, R.M. Russel, X.-D. Wang. The biochemical characterization of ferret carotene-9′, 10′-monooxygenase catalyzing cleavage of carotenoids in vitro and in vivo. J Biol Chem, 281 (2006), 19327–19338.

[15] M. Gajic, S. Zaripheh, F.R. Sun, J.W. Erdman Jr. Apo-8′-lycopenal and apo-12′-lycopenal are metabolic products of lycopene in rat liver. J Nutr, 136 (2006), 1552–1557.

[16] E. Reynaud, G. Aydemir, R. Ruhl, O. Dangles, C. Caris-Veyrat. Organic synthesis of new putative lycopene metabolites and preliminary investigation of their cell-signaling effects. J Agri Food Chem, 59 (2011), 1457–63.

[17] R.A. Bone, J.T. Landrum, L.H. Guerra, C.A. Ruiz. Lutein and zeaxanthin dietary supplements raise macular pigment density and serum concentrations of these carotenoids in humans. J Nutr, 133 (2003), 992–8.

[18] L. Yonekura, M. Kobayashi, M. Terasaki, A. Nagao. Keto-carotenoids are the major metabolites of dietary lutein and fucoxanthin in mouse tissues. J Nutr, 140 (2010), 1824–1831.

[19] F. Khachik, F.F. de Moura, D.Y. Zhao, C.P. Aebischer, P.S. Bernstein. Transformations of selected carotenoids in plasma, liver and ocular tissues of humans and in nonprimate animal models. Invest Ophthalmol Visual Sci, 43 (2002), 3383–3392.

[20] R.A. Bone, J.T. Landrum, G.W. Hime, A. Cains, J. Zamor. Stereochemistry of the human macular carotenoids. Invest Ophthalmol Visual Sci, 34 (1993), 2033–2040.

[21] J.R. Mein, G.G. Dolnikowski, H. Ernst, R.M. Russell, X.D. Wang. Enzymatic formation of apo-carotenoids from the xanthophyll carotenoids lutein, zeaxanthin and β-cryptoxanthin by ferret carotene-9′,10′-monooxygenase. Arch Biochem Biophy, 506 (2011), 109–121.

[22] T. Sugawara, V. Baskaran, W. Tuzuki, A. Nagao. Brown alagae Fucoxanthin in hydrolyzed to fucoxanthinol during absorbtion by Caco2 human intestinal cells and mice. J Nutr, 132 (2002), 946–951.

[23] A. Asai, T. Sugawara, H. Ono, A. Nagao. Biotransformation of fucoxanthinol into amarouciaxanthin A in mice and HepG2 cells: formation and cytotoxicity of fucoxanthin metabolites. Drug Metabol Disposition, 32 (2004), 205–211.

[24] E. Wolz, H. Liechti, B. Notter, G. Oesterhelt, A. Kistler. Characterization of metabolites of astaxanthin in primary cultures of rat hepatocytes. Drug Metabol Disposition, 27 (1999), 456–462.

[25] A. Kistler, H. Liechti, L. Pichard, E. Wolz, G. Oesterhelt, A. Hayes, P. Maurel. Metabolism and CYP-inducer properties of astaxanthin in man and primary human hepatocytes. Arch Toxicol, 75 (2002), 665–75.

[26] Y. Yang, B. Kim, J. Young Lee. Astaxanthin structure, metabolism, and health benefits. J Human Nutr Food Sci, 1003 (2013), 1–11.

[27] J.K. Tyczkowski, B. Yagen, P.B. Hamilton. Metabolism of canthaxanthin, a red diketo-carotenoid, by chickens. Poultry Science, 67 (1988), 787–93.

[28] V. Gokmen, S. Bahceci, J. Acar. Liquid chromatographic method for the determination of chlorophylls, carotenoids, and their derivatives in fresh and processed vegetables. J Liq Chromatogr Related Tech, 25 (2002), 1201–1213.

[29] Q. Su, K.G. Rowley, N.D.H. Balazs. Carotenoids: separation methods applicable to biological samples. J Chromatogr B, 781 (2002), 393–418.

[30] K.S. Epler, R.G. Ziegler, N.E. Craft. Liquid chromatographic method for the determination of carotenoids, retinoids and tocopherols in human serum and in food. J Chromatogr, 619 (1993), 37–48.

[31] R.A. Bone, J.T. Landrum, Z. Dixon, Y. Chen, C.M. Llerena. Lutein and zeaxanthin in the eyes, serum and diet of human subjects. Exp Eye Res, 71 (2000), 239–245.

[32] S.P. Stratton, W.H. Schaefer, D.C. Liebler. Isolation and identification of singlet oxygen oxidation products of β-carotene. Chem Res Toxicol, 6 (1993), 542–547.

[33] R. Wyss, F. Bucheli. Determination of endogenous levels of 13-cis-retinoic acid (isotretinoin), all-trans-retinoic acid (tretinoin) and their 4-oxo metabolites in human and animal plasma by high-performance liquid chromatography with automated column switching and ultraviolet detection. J Chromatogr B, 700 (1997), 31–47.

[34] W.G. Siems, O. Sommerburg, J.S. Hurst, F.J. van Kuijk. Carotenoid oxidative degradation products inhibit Na+-K+-ATPase. Free Rad Res, 33 (2000), 427–435.

[35] S.J. Kim, E. Nara, H. Kobayashi, J. Terao, A. Nagao. Formation of cleavage products by autoxidation of lycopene. Lipids, 36 (2001), 191–199.

[36] P.S. Bernstein, F. Khachik, L.S. Carvalho, G.J. Muir, D.Y. Zhao, N.B. Katz. Identification and quantitation of carotenoids and their metabolites in the tissues of the human eye. Exp Eye Res, 72 (2001), 215–223.

[37] M. Dachtler, T. Glaser, K. Kohler, K. Abert. Combined HPLC-MS and HPLC-NMR online coupling for the separation and determination of lutein and Zeaxanthin stereoisomers in spinach and in retina. Anal Chem, 73 (2001), 667–674.

[38] O. Aust, N. Ale-Agha, L. Zhang, H. Wollersen, H. Sies, W. Stahl. Lycopene oxidation product enhances gap junctional communication. Food Chem Toxicol, 41 (2003), 1399–1407.

[39] O. Sommerburg, C.D. Langhans, J. Arnhold, M. Leichsenring, C. Salerno, C. Crifò, G.F. Hoffmann, K.M. Debatin, W.G. Siems. β-carotene cleavage products after oxidation mediated by hypochlorous acid -a model for neutrophil-derived degradation. Free Rad Biol Med, 35 (2003) 1480–90.

[40] P.D. Fraser, M.A. Eugenia, Enfissi, M. Goodfellow, T. Eguchi, P.M. Bramley. Metabolite profiling of plant carotenoids using the matrix-assisted laser desorption ionization time-of-flight mass spectrometry. The Plant Journal, 49 (2007), 552–564.

[41] P. Crupi, R.A. Milella, D. Antonacci. Simultaneous HPLC-DAD-MS (ESI+) determination of structural and geometrical isomers of carotenoids in mature grapes. J Mass Spectrom, 45 (2010), 971–980.

[42] M. Rivera, P. Christou, R. Canela-Garayoa. Identification of carotenoids using Mass Spectrometry. Mass Spectrom Rev, 33 (2014), 353–372.

[43] P. Bhosale, D.Y. Zhao, B. Serban, P.S. Bernstein. Identification of 3-methoxyzeaxanthin as a novel age-related carotenoid metabolite in the human. Invest Ophthalmol Vis Sci, 48 (2007), 1435–1440.

[44] R. Lakshminarayana, G. Aruna, R.K. Sangeetha, N. Bhaskar, S. Divakar, V. Baskaran. Possible degradation/biotransformation of lutein in vitro and in vivo: isolation and structural elucidation of lutein metabolites by HPLC and LC-MS (atmospheric pressure chemical ionization). Free Rad Biol Med, 45 (2008), 982–993.

[45] R.K. Sangeetha, N. Bhaskar, V. Baskaran. Comparative effects of β-carotene and fucoxanthin on retinol deficiency induced oxidative stress in rats. Mol Cell Biochem, 331 (2009), 59–67.

[46] R.B. van Breemen, D. Nikolic, X. Xu, Y. Xiong, M. van Lieshout, C.E. West, A.B. Schilling. Development of a method for quantitation of retinol and retinyl palmitate in human serum using high-performance liquid chromatography-atmospheric pressure chemical ionization-mass spectrometry. J Chromatogr A, 79 (1998), 245–51.

[47] D. Zhu, Y. Wang, Y. Pang, A. Liu, J. Guo, C.A. Bouwman, C.E. West, R.B. van Breemen. Quantitative analyses of β-carotene and retinol in serum and feces in support of clinical bioavailability studies. Rap Comm Mass Spectrom, 20 (2006), 2427–32.

[48] M.W. Schäffer, S.S. Roy, S. Mukherjee, D. Nohr, M. Wolter, H.K. Biesalski, D.E. Ong, S.K. Das. Qualitative and quantitative analysis of retinol, retinyl esters, tocopherols and selected carotenoids out of various internal organs from different species by HPLC. Anal Methods, 2 (2010), 1320–1332.

[49] P. Würtinger, H. Oberacher. Evaluation of the performance of a tandem mass spectral library with mass spectral data extracted from literature. Drug Test Anal, 4 (2012), 235–241.

[50] D. Giuffrida, A. Pintea, P. Dugo, G. Torre, R.M. Pop, L. Mondello. Determination of carotenoids and their esters in fruits of sea buckthorn (Hippophae rhamnoides L.) by HPLC-DAD-APCI-MS. Phytochem Anal, 23 (2012), 267–73.

[51] S. Rivera, F. Vilaró, R. Canela. Determination of carotenoids by liquid chromatography/mass spectrometry: effect of several dopants. Anal Bioanal Chem, 400 (2011), 1339–46.

[52] J.K. Prasain, R. Moore, J.S. Hurst, S. Barnes, K.J. van Kuijk. Electrospray tandem mass spectrometric analysis of zeaxanthin and its oxidation products. J Mass Spectrom, 40 (2005), 916–923.

[53] F. Granado-Lorencio, C. Herrero-Barbudo, I. Blanco-Navarro, B. Pérez-Sacristán Suitability of ultra-high performance liquid chromatography for the determination of fat-soluble nutritional status (vitamins A, E, D, and individual carotenoids). Anal Bioanal Chem, 39 (2010), 1389–93.

[54] A.D. Rius, J. Eras, A.M. Vall, F. Vilaró, M. Balcells, R.C. Garayoa. Ultra performance liquid chromatography analysis to study the changes in the carotenoid profile of commercial monovarietal fruit juices. J Chromatogr A 1331, 28 (2014), 90–99.

[55] B. Li, H. Zhao, J. Liu, W. Liu, S. Fan, G. Wu, R. Zhao. Application of ultra-high performance supercritical fluid chromatography for the determination of carotenoids in dietary supplements. J Chromatogr A, 1425 (2015), 28–292.

[56] B.P. Arathi, P.R. Sowmya, K. Vijay, P. Dilshad, B. Saikat, P. Gopal, R. Lakshminarayana. An improved method of UPLC-PDA-MS/MS analysis of lycopene isomers. Food Anal Method, 8 (2014), 1962–1969.

[57] L. Dong, H. Shion, R.G. Davis, B. Terry-Penak, J. Castro-Perez, R.B. van Breemen. Collision cross-section determination and tandem mass spectrometric analysis of isomeric carotenoids using electrospray ion mobility time-of-flight mass spectrometry. Anal Chem, 82 (2010), 9014–9021.

[58] C.M. Raphael, C.W. John. Identification of products from canthaxanthin oxidation. Food Chem,197 (2016), 836–840.

[59] T. Kind, O. Fiehn Metabolomic database annotations via query of elemental compositions: mass accuracy is insufficient even at less than 1 ppm. BMC Bioinformatics, 7 (2006), 234.

[60] L. Van Meulebroek, J. Vanden Bussche, K. Steppe, L. Vanhaecke. Unravelling the phytohormonal status and carotenoid profile of tomato plants by using metabolomics, including generic extraction and UHPLC-Orbitrap-MS. Anal Bioanal Chem, 406 (2014), 2613–2626.

[61] F. Cacciola, P. Donato, D. Giuffrida, G. Torre, P. Dugo, L. Mondello. Ultra high pressure in the second dimension of a comprehensive two-dimensional liquid chromatographic system for carotenoid separation in red chilli peppers. J Chromatogr A, 1255 (2012), 244–251.

[62] C. Tode, T. Maoka, M. Sugiura. Application of LC-NMR to Analysis of carotenoids in foods. J Sep Sci, 32 (2009), 3659–3663.

[63] M.H. Zile, M.E. Cullum, R.U. Simpson, A.B. Barua, D.A. Swartz. Induction of differentiation of human promyelocytic leukemia cell line HL-60 by retinoyl glucuronide, a biologically active metabolite of vitamin A. PNAS, 84 (1987), 2208–12.

[64] J.Y. Winum, M. Kamal, H. Defacque, T. Commes, C. Chavis, M. Lucas, J. Marti, J.L. Montero. Synthesis and biological activities of higher homologues of retinoic acid. Farmaco, 52 (1997), 39–42.

[65] T. Suzuki, M. Matsui, A. Murayama. Biological activity of (all-E)-β-apo-12′-carotenoic acid and the geometrical isomers on human acute promyelocytic leukemia cell line HL-60. J Nutr Sci Vitaminol, 41(1995), 575–85.

[66] X. Hu, K.M. White, N.E. Jacobsen, D.J. Mangelsdorf, L.M. Canfield. Inhibition of growth and cholesterol synthesis in breast cancer cells by oxidation products of β-carotene. J Nutr Biochem, 9 (1998), 567–574.

[67] T. Kawada, Y. Kamei, A. Fujitha, Y. Hida, N. Takahashi, E. Sugimotot, T. Fushiki. Carotenoids and retinoids as suppressors on adipocyte differentiation via nuclear receptors. Biofactors, 13 (2000), 103–9.

[68] E.C. Tibaduiza, J.C. Fleet, R.M. Russell, N.I. Krinsky. Excentric cleavage products of β-carotene inhibit estrogen receptor positive and negative breast tumor cell growth in vitro and inhibit activator protein-1-mediated transcriptional activation. J Nutr, 132 (2002), 1368–75.

[69] R. Rühl. Induction of PXR-mediated metabolism by β-carotene. Biochimica et Biophysica Acta (BBA) - Mol Basis of Disease, 1740 (2005), 162–169.

[70] O. Ziouzenkova, G. Orasanu, G. Sukhova, E. Lau, J.P. Berger, G. Tang, N.I. Krinsky, G.G. Dolnikowski, J. Plutzky. Asymmetric cleavage of β-carotene yields a transcriptional repressor of retinoid X receptor and peroxisome proliferator-activated receptor responses. Mol Endocrinol, 21 (2007), 77–88.

[71] A. Eroglu, D.P. Hruszkewycz, C. dela Sena, S. Narayanasamy, K.M. Ried, R.E. Kopec, S.J. Schwartz, R.W. Curley Jr, E.H. Harrison. Naturally occurring eccentric cleavage products of provitamin A β-carotene function as antagonists of retinoic acid receptors. J Biol Chem, 287 (2012), 15886–95.

[72] E.N. Story, R.E. Kopec, S.J. Schwartz, G.K. Harris. An update on the health effects of tomato lycopene. Ann Rev Food Sci Technol, 1 (2010), 189–210.

[73] N.I. Ivanov, S.P. Cowell, P. Brown, P.S. Rennie, E.S. Guns, M.E. Cox. Lycopene differentially induces quiescence and apoptosis in androgen-responsive and -independent prostate cancer cell lines. Clin Nutr, 26 (2007), 252–263.

[74] E.S. Hwang, P.E. Bowen. Effects of lycopene and tomato paste extracts on DNA and lipid oxidation in LNCaP human prostate cancer cells. Biofactors, 23, (2005), 97–105.

[75] L. Tang, T. Jin, X. Zeng, J.S. Wang. Lycopene inhibits the growth of human androgen-independent prostate cancer cells in vitro and in BALB/c nude mice. J Nutr, 135 (2005), 287–290.

[76] B.L. Lindshield, K.C. Adams, J.W. Erdman Jr. Lycopenoids: Are lycopene metabolites bioactive. Arch Biochem Biophy, 458 (2007), 136–140.

[77] J. Mein, X.D. Wang. Oxidative metabolites of lycopene and their biological functions. In: Landrum JT (ed.). Boca Raton, FL: Taylor and Francis Group, (2010), 417–35.

[78] X.D. Wang. Lycopene metabolism and its biological significance. Am J Clin Nutr, 96 (2012), 1214–1222.

[79] C.C. Veyrat, A. Schmid, M. Carail, V. Böhm. Cleavage products of lycopene produced by in vitro oxidations: characterization and mechanisms of formation. J Agri Food Chem, 51 (2003), 7318–7325.

[80] E. Nara, H. Hayashi, M. Kotake, K. Miyashita, A. Nagao. Acyclic carotenoids and their oxidation mixtures inhibit the growth of HL-60 human promyelocytic leukemia cells. Nutr Cancer, 39 (2001), 273–283.

[81] H. Zhang, E. Kotake-Nara, H. Ono, A. Nagao. A novel cleavage product formed by autoxidation of lycopene induces apoptosis in HL-60 cells. Free Rad Biol Med, 35 (2003), 1653–1663.

[82] N.A. Ford, A.C. Elsen, K. Zuniga, B.L. Lindshield. Lycopene and Apo-12′-Lycopenal Reduce Cell Proliferation and Alter Cell Cycle Progression in Human Prostate Cancer Cells. Nutr Cancer, 63(2011), 256–263.

[83] J. Chung, K. Koo, F. Lian, K.Q. Hu, H. Ernst, X.D. Wang. Apo-10′-Lycopenoic Acid, a Lycopene Metabolite, Increases Sirtuin 1 mRNA and Protein Levels and Decreases Hepatic Fat Accumulation in ob/ob Mice. J Nutr, 142 (2012), 405–410.

[84] R. Lakshminarayana, V.U. Sathish, M.S. Dharmesh, V. Baskaran. Antioxidant and cytotoxic effect of oxidized lutein in human cervical carcinoma cells (HeLa). Food Chem Toxicol, 48 (2010), 1811–1816.

[85] A.A. Woodall, G. Britton, M.J. Jackson. Carotenoids and protection of phospholipids in solution or liposomes against oxidation by peroxyl radicals: relationship between carotenoid structure and protective ability. Biochim Biophys Acta, 1336 (1997), 575–586.

[86] R.K. Sangeetha, V. Baskaran. Retinol-deficient rats can convert a pharmacological dose of astaxanthin to retinol: antioxidant potential of astaxanthin, lutein, and β-carotene. Canadian J Physiol Pharmacol, 88 (2010), 977–85.

[87] K.N. Kim, S.J. Heo, S.M. Kang, G. Ahn, Y.J. Jeon. Fucoxanthin induces apoptosis in human leukemia HL-60 cells through a ROS-mediated Bcl-xL pathway. Toxicol In Vitro, 24 (2010), 1648–1654.

[88] M.N. Woo, S.M. Jeon, Y.C. Shin, M.K. Lee, M.A. Kang, M.S. Choi. Anti-obese property of fucoxanthin is partly mediated by altering lipid-regulating enzymes and uncoupling proteins of visceral adipose tissue in mice. Mol Nutr Food Res, 53 (2009), 1603–1611.

[89] A. Rwigemera, J. Mamelona, L.J. Martin. Comparative effects between fucoxanthinol and its precursor fucoxanthin on viability and apoptosis of breast cancer cell lines MCF-7 and MDA-MB-231. Anticancer Res, 35 (2015), 207–219.

[90] L.J. Martin. Fucoxanthin and its metabolite fucoxanthinol in cancer prevention and treatment. Marine Drugs, 13 (2015), 4784–4798.

[91] H. Maeda, M. Hosokawa, T. Sashima, N. Takahashi, T. Kawada, K. Miyashita. Fucoxanthin and its metabolite, fucoxanthinol, suppress adipocyte differentiation in 3T3-L1 cells. International J Mol Med, 18 (2006), 147–152.

[92] A. Ben-Dor, A. Nahum, M. Danilenko, Y. Giat, W. Stahl, H.D. Martin, T. Emmerich, N. Noy, J. Levy, Y. Sharoni. Effects of acyclo-retinoic acid and lycopene on activation of the retinoic acid receptor and proliferation of mammary cancer cells. Arch Biochem Biophy, 391 (2001), 295–302.

[93] P.K. Duitsman, A.B. Barua, B. Becke, J.A. Olson. Effects of epoxycarotenoids, β-carotene, and retinoic acid on the differentiation and viability of the leukemia cell line NB4 in vitro. International J Vitaminol Nutr Res, 69 (1999), 303–308.

[94] I. Konishi, M. Hosokawa, T. Sashima, H. Kobayashi, K. Miyashita. Halocynthiaxanthin and fucoxanthinol isolated from Halocynthia roretzi induce apoptosis in human leukemia, breast and colon cancer cells. Comp Biochem Physiol,142 (2006), 53–59.

[95] K. Yamamoto, C. Ishikawa, H. Katano, T. Yasumoto, N. Mori. Fucoxanthin and its deacetylated product, fucoxanthinol, induce apoptosis of primary effusion lymphomas. Cancer Letter, 300 (2011), 225–234.

[96] M. Hanusch, W. Stahl, W.A. Schulz, H. Sies. Induction of gap junctional communication by 4-oxoretinoic acid generated from its precursor canthaxanthin. Arch Biochem Biophy, 17 (1995), 423–428.

Chapter 3

Metabolomic and Proteomic Analysis of the Mesenchymal Stem Cells' Secretome

Galya Ivanova, Tiago Pereira, Ana Rita Caseiro,
Petia Georgieva and Ana Colette Maurício

Additional information is available at the end of the chapter

http://dx.doi.org/10.5772/66101

Abstract

Mesenchymal stem cells (MSCs) are multipotent stromal cells with a strong potential in human regenerative medicine due to their ability to renew themselves and differentiate into various specialized cell types under certain physiological or experimental conditions. MSCs secrete a broad spectrum of autocrine and paracrine factors (MSCs' secretome) that could exert significant effects on cells in their vicinity. MSCs have been clinically tested and have displayed a great potential in the treatment of bone/cartilage fractures and disorders, diabetes, cardiovascular diseases and immune, neurodegenerative and inflammatory diseases. The therapeutic efficacy of MSCs was initially attributed to their multipotent character and ability to engraft and differentiate at the site of injury. However, in recent years, it has been revealed that either undifferentiated or differentiated MSCs' secretome plays an important role in the therapeutic potential of MSCs. The deciphering of the composition of MSCs' secretome through proteomic and metabolic analyses and implementation of certain advanced analytical (nuclear magnetic resonance (NMR) spectroscopy, mass spectrometry (MS), chromatography, etc.) and immunological methods could contribute to the understanding of the mechanisms underlying the therapeutic effects of MSCs.

Keywords: secretome, mesenchymal stem cells, proteomics, metabolomics, umbilical cord blood plasma, growth factors

1. Introduction

Mesenchymal stem cells (MSCs) are one of the most promising types of stem cells for cell-based therapies. The continuously increasing interest in the therapeutic application of MSCs to various diseases can be linked to their capacity to self-renew and differentiate into several cell types, to secret soluble factors with paracrine actions and their immunosuppressive and immunomodulatory properties [1–5]. In addition, MSCs can be easily isolated from a wide range of ethically acceptable adult tissues and demonstrate a significant capacity for *ex vivo* expansion [6, 7].

Mesenchymal stem cells (MSCs) were first isolated and described by Friedenstein et al. in the early 1970s as a population of multipotent cells found in the bone marrow that can differentiate into osteocytes, chondrocytes, adipocytes and myoblasts [8, 9]. Currently, MSCs can be found and isolated from various foetal and adult tissues including bone marrow, adipose tissue, skeletal muscle, umbilical cord blood (UCB), umbilical cord tissue or matrix (UCT, Wharton's jelly), peripheral blood, dental pulp and amniotic fluid [10–12]. Despite their presence in different tissue sources, the isolated MSCs exhibit a similar characteristic phenotype [2] with some additional features that reflect their tissue of origin [13].

MSCs are able to renew themselves and differentiate into various specialized cell types such as bone, cartilage, muscle and fat cells under certain physiological or experimental conditions and may serve as a renewable source of cell and tissue replacements [1, 2]. According to the definition agreed by the International Society for Cellular Therapy (ISCT) in 2006, mesenchymal stem cells are characterized by (a) their capacity to adhere to plastic; (b) their expression of specific surface markers, namely, CD73, CD90 and CD105, and no expression of CD14, CD19, CD34, CD45 and HLA-DR. Additionally, according to the ISCT, MSCs are able to undergo trilineage differentiation into adipocytes, chondrocytes and osteoblasts [2].

Besides their multipotency, MSCs possess unique characteristics, such as the ability to migrate to sites of injury, inflammation or cancer, the capacity to secrete a broad spectrum of autocrine/paracrine factors that could exert significant effects on cells in their vicinity and the ability to modulate inflammatory and regenerative processes and the immune system. MSCs have been clinically tested for treatment of a variety of pathologies and diseases such as brain paralysis, cardiovascular diseases and myocardial infarction, type I diabetes, multiple sclerosis, Crohn's disease, bone fractures, graft versus host disease (GVHD) in bone marrow transplantation, osteoarthritis and rheumatoid arthritis [3, 14, 15].

The therapeutic effects of MSCs were initially attributed to the ability to migrate to sites of injury and inflammation, engraft into the damaged tissues and differentiate into specialised cell types. However, many studies provide strong evidences that the therapeutic efficacy of transplanted MSCs is not dependent on the physical proximity of the transplanted cells to the damaged tissue and propose that MSCs exert their therapeutic activity through MSCs' secretome [16, 17]. Although these studies, and many others, indicate the potency of MSC-secreted factors in mediating tissue repair and regeneration, the mechanisms of MSCs' action are not completely clear. The scientific community has tried to understand MSCs' biological mechanisms of action considering their capacity to secret soluble factors with paracrine actions and unveil their potential in cell therapy and regenerative medicine.

Proteomics and metabolomics are omics techniques that employ state of the art analytical instrumentation in conjunction with pattern recognition techniques for comprehensive and simultaneous systematic analysis of biological systems and monitoring of cellular and systemic proteomic and metabolic fluctuations in response to xenobiotic exposure, environmental factors, physiological stimuli and genetic modifications [18–23].

The studies of the composition of secretion products of MSCs through proteomic and metabolic analyses could contribute to the understanding of the mechanisms underlying the therapeutic effects of MSCs. Currently, certain advanced analytical (nuclear magnetic resonance (NMR) spectroscopy, mass spectrometry (MS), chromatography, immunological assays, etc.) and chemometric techniques exist for proteomic and metabolic analyses of bio-fluids, which can be applied in MSC secretomics [20, 22–24].

2. MSCs' conditioned medium and secreted products

The paracrine effect of MSCs and their ability to synthesise and secrete a broad spectrum of growth factors, chemokines and cytokines that could exert significant effects on cells in their vicinity were first described by Haynesworth et al. [25]. This study was followed by other publications in the scientific literature describing therapeutic effects of these secreted factors [26, 27].

The MSCs' secreted factors are principal molecules for intercellular communication involved in most physiological processes, such as cell signalling, differentiation, invasion, metastasis, cell adhesion and binding, angiogenesis and apoptosis [18]. Growth factors, extracellular matrix proteins and extracellular matrix remodelling enzymes, pro-inflammatory, anti-inflammatory and pleiotropic cytokines, chemokines and angiogenic factors have been recently identified in stem cell secretomes including MSCs' one [19].

The compounds synthesised in the process of the proliferation and differentiation of MSCs could be secreted from the cells either constitutively or in the regulated manner and determine the composition of the extracellular environment [28]. Cell treatments *in vitro* could initiate alterations in the amount of particular factors secreted in the culture media [29, 30]. The secreted molecules could act as a possible replacement of stem cells for therapeutic applications that allow precise dosing, low biological variability and overcome of several stem cell related issues including cell origin and immunocompatibility [20]. The therapeutic application of MSCs' secretome could reveal new safe and effective strategies with predictable outcomes as an alternative to the cell therapy.

2.1. Preparation of MSCs' conditioned medium samples

MSCs' secretome studies involve a number of defined stages: (i) cell isolation and characterisation; (ii) cell culture in an appropriate culture medium; (iii) cell expansion and differentiation and (iv) collection of conditioned media.

The preparation of MSCs' conditioned medium has been performed by our research group and published elsewhere and here it is resumed and described [31]. Different batches of human MSCs from Wharton's jelly umbilical cord cryopreserved cells are *in vitro* cultured and maintained in a humidified atmosphere with 5% CO_2 at 37°C. This process can be applied to previously isolated, expanded and cryopreserved MSCs from a diversity of tissues, with the advantage of a higher number of MSCs obtained in a shorter culture time, not dependent on donor's availability and ethic committee's authorization. During expansion, the cells become long spindle-shaped, and their phenotype is confirmed by flow cytometry analysis for a comprehensive panel of markers, such as platelet endothelial cell adhesion molecule-1 (PECAM-1, CD31), homing cell adhesion molecule (HCAM, CD44), CD45 and Endoglin (CD105). Characterization is also performed with the following antibodies and their respective isotypes: PE anti-human CD105; APC anti-human CD73; PE anti-human CD90; PerCP/Cy5.5 anti-human CD45: FITC anti-human CD34; PerCP/Cy5.5 anti-human CD14; Pacific Blue anti-human CD19 and pacific-blue anti-human HLA-DR. The karyotype of the MSCs should also be determined to ensure that no structural alterations are found, which demonstrate the absence of neoplastic characteristics in these cells, as well as chromosomal stability of the somatic and sexual chromosomes due to the cell culture procedures. Two mesenchymal stem cell media have been tested for the conditioning, namely, PromoCell® medium (LabClinics, Promocell, C-28010, so-called commercial medium further on) and Dulbecco's Modified Eagle Medium (DMEM, Gibco®) supplemented with 10% of foetal bovine serum (FBS, Gibco®), 2 mM glutamine (sigma), 100 U/ml of penicillin and 100 μg/ml of streptomycin (Sigma). Conditioned medium (CM) is normally collected from passage 4 (P4) MSCs. To obtain the desired CM, 4000 cell/cm^2 are plated and allowed to grow until reaching a minimum of 80% confluence. At this stage, the commercial medium is removed from the T-flasks, and after five washing cycles with Dulbecco's Phosphate Buffered Saline 1× (DPBS) without calcium (Ca^{2+}) and magnesium (Mg^{2+}) (Gibco®), Dulbecco's Modified Eagle Medium/Nutrient Mixture (DMEM, Gibco®) supplemented with 100 U/ml of penicillin and 100 μg/ml of streptomycin (Sigma) is added. The cell culture T-flasks are maintained in a humidified atmosphere with 5% CO_2 at 37°C, allowing the adherent cells to be in contact with a serum-free basal medium. The culture medium, added after the confluence is reached, is not renewed and it is collected at different time points (24 and 48 h). For conditioning with Dulbecco's Modified Eagle Medium/Nutrient Mixture (DMEM, Gibco®) supplemented with 100 U/ml of penicillin and 100 μg/ml of streptomycin (Sigma), samples of the medium are also collected at 24 and 48 h. Upon collection, the CM is frozen at –20°C, being later on thawed on the day of the experiments. ^{1}H-NMR spectra and Multiplexing LASER Bead analysis were acquired from both 24 and 48 h samples.

3. Human umbilical cord blood plasma as a potential supplement for MSCs' culture and conditioning

In search of a suitable alternative (and animal-component-free) culturing medium for the referred MSCs, we employed similar analytical techniques in order to better characterize hUCBP and assess its viability as a substitute for standard usage in FBS. The utilization of FBS in cellular culture has a series of ethical, technical and biological issues that made challenging its most extended usage in clinical setting. Since exogenous protein and other factors' supplementation is still a facilitating point in MSCs' expansion (as chemically defined media are mostly expensive and occasionally underperforming), research has focused on finding suitable replacements for the FBS [31–34].

3.1. Preparation of hUCBP samples

These samples were collected from different consenting donors that clinically evaluated according to the Portuguese law 12/2009 (Diário da República, lei 12/2009 de 26 de Março de 2009) and analysed by flow cytometry and for microbiological contamination for aerobic and anaerobic microorganisms and fungi.

Umbilical cord blood (UCB) was collected from the umbilical vein by gravity into a 150 ml volume simple bag (reference 1385.13, Suru, Portugal), containing 21 ml of citrate-phosphate-dextrose (CPD), and stored at 4 ± 2°C until processing for cryopreservation. UCB samples were transported to the laboratory at refrigerated temperatures ranging between 4 and 22°C, within 48 h after collection. The collected UCB is subjected to a volume reduction process using AXP system® (thermogenesis). During the two-step centrifugation, whole blood is separated into three layers that are delivered into a red blood cell (RBC) bag and a freezing bag. Plasma remains in the processing bag which is also known as plasma bag. Samples of the UCB are taken by sampling pillows integrated within the kit for flow cytometry analysis. The total nucleated cell (TNC) count, CD34+ cell counts, CD34+ cell viability and leucocyte (CD45+) viability are determined by samples obtained from the UCB before volume reduction. Microbiological controls are usually performed after volume reduction and before cryopreservation and tested for microbiological contamination using an automated blood culture system (BacT/ALERT®, bioMérieux) at 35°C for 14 days. TNC and the number of white blood cells (WBC) are counted with a haematology autoanalyser (Ac T diff2™, Beckman Coulter, Inc.). The CD34+ cell number and the CD34+ viability are quantified by flow cytometry (BD FACSCalibur™ 3 CA Becton Dickinson, BD Biosciences); the software for acquisition and analysis was BD CellQuest™ and BDCellQuest™ Pro Templates, respectively. The clusters of differentiation (CD) were used to enumerate the total number of CD34+ cells and the total number of leucocytes (CD45+), and the 7-amino-actinomycin D (7AAD) nucleic acid dye was used for viability measure (BD Stem Cell Enumeration kit, Becton Dickinson, BD Biosciences), according to the manufacture's protocol. The BD stem cell enumeration simultaneously enumerates the total viable dual-positive (CD45+/CD34+) haematopoietic stem cells in absolute counts of CD34+ cells per μl and the percentage (%) of viable leucocytes (CD45+) that are CD34 positive (CD34+).

The sample collection, sample storage and sample preparation are very important steps in proteomics and metabolomics. If the samples are not collected properly or the samples are not stored or processed uniformly, the metabolomics data generated from these samples could be invalid. Hence, the sample collection, storage and processing procedures are critical for conducting successful metabolomics studies.

4. Proteomic techniques

Proteomics is the omics technique that focuses on proteins and their functions in biological systems. Proteins are involved in all processes of living organisms and possess a complex structure, interactions, dynamics in concentration, degradation and/or modification which are the key factors defining the behaviour of biological systems. Therefore, the main goal of proteomics is the identification, quantification and characterisation of the protein content of biological samples, such as organs, tissues, cells and biological fluids [22].

Well-established methodologies exist for proteomic analysis of bio-fluids that include separation and protein identification techniques, and chemometric methods for data interpretation and visualization [18, 20, 24]. The extensive development of advanced analytical techniques and instrumentation in the last decade has enabled the proteomic analysis of cell secretome and several profiles from different cell types, body fluids and physiological conditions that have been studied and established [20, 35]. The continuously increasing interest in secretome proteomics has been raised due to the pivotal role of these secreted proteins in all physiological processes in living systems, including physiological, pathophysiological and genetic transformations, degenerative processes, disease conditions and progression [23, 24, 28].

4.1. Proteomic analysis of MSCs' secretome

The proteomic profiling of biological systems, including cell secretomes is mainly centred on mass spectrometry (MS), chromatography (liquid chromatography, LC), immunological assays and chemometric techniques [20, 24]. *In vivo* and *ex vivo* studies of MSCs' secretome encounter significant technical difficulties due to the low abundance of secreted proteins relative to total proteins in tissue/bio-fluid, the presence of different cell types including the cells of interest, endothelial cells and fibroblast cells, possibility of contamination by blood and the difficulty of accessing the extracellular medium in a tissue section [20, 23]. Therefore, *in vitro* strategies have been developed to probe the secretome under physiological conditions, where culture medium is conditioned by cells for a certain period of time followed by collection, preparation and processing of the medium for proteomic analysis. The proteomic analysis of MSCs' secreted products involves MSCs' isolation and characterisation; culture medium preparation; MSCs' conditioning; isolation of the conditioned medium; implementation of appropriate proteomic analytical techniques and protocols and data analysis. Proteomic studies of MSCs' secretome are currently performed under *in vitro* conditions as shown in **Figure 1** [20, 23].

Proteomic analysis based on LC, MS, LC-MS, immunological assays and bioinformatics has already been applied in the studies of protein and peptide separation, identification and quantification, screening of posttranslational modifications and characterisation of secreted products [20, 22, 23, 36, 37]. Proteomic studies performed on MSCs' secretome have been recently revised considering the cell origin, conditioning protocols and the analytical techniques used [20]. The studies of the secreted products of MSCs isolated from various foetal and adult tissues including bone marrow, adipose tissue, skeletal muscle, umbilical cord blood and umbilical cord tissue have been performed. Proteomic profiles of secreted products obtained from culturing of a wide range of human and animal MSCs have been established through implementation of proteomic analytical techniques such as MS, LC, LC/MS, immunological assays and bioinformatics [20, 38–42].

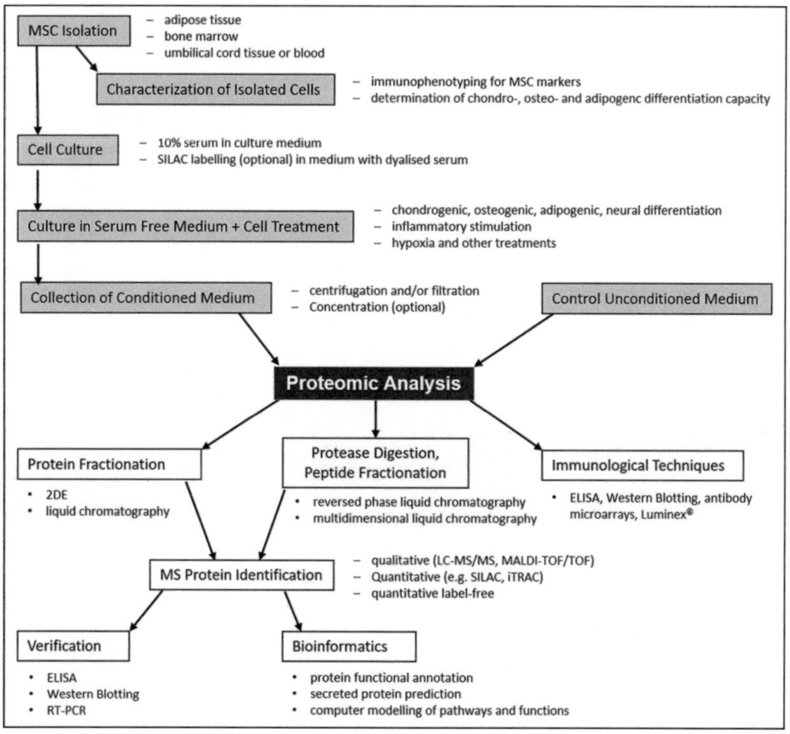

Figure 1. Schematic representation of the steps followed in a proteomic analysis of conditioned medium obtained from MSCs' *in vitro* expansion.

We have recently disclosed proteomic study of several MSCs' conditioned media based on Multiplexing LASER Bead Technology which enable the simultaneous testing of numerous analytic categories such as cytokines, chemokines and growth factors in a single assay [31]. Human Primary Cytokine Array/Chemokine Array 41-Plex Panel (Eve Technologies, Calgary, Alberta, Canada) has been performed to analyse the conditioned media, including the following cytokines, chemokines and growth factors: epidermal growth factor (EGF), eotaxin-1, fibroblast growth factor 2 (FGF-2), fms-related tyrosine kinase 3 ligand (Flt-3L), fractalkine, granulocyte colony-stimulating factor (G-CSF), granulocyte macrophage colony-stimulating factor (GM-CSF), GRO(pan), interferon- alpha 2 (IFNα2), interferon-gamma (IFNγ), several interleukins (IL-1α, IL-1β, IL-1ra, IL-2, IL-3, IL-4, IL-5, IL-6, IL-7, IL-8, IL-9, IL-10, IL-12 (p40), IL-12 (p70), IL-13, IL-15, IL-17A), interferon gamma-induced protein 10 (IP-10), monocyte chemotactic protein-1 (MCP-1), monocyte chemotactic protein-3 (MCP-3), macrophage-derived chemokine (MDC), macrophage inflammatory protein-1 alpha (MIP-1α), macrophage inflammatory protein-1 beta (MIP-1β), platelet-derived growth factor-AA (PDGF-AA), platelet-derived growth factor-AB/BB (PDGF-AB/BB), chemokin (C-C motif) ligand 5 (RANTES or CCL5), soluble CD40 ligand (sCD40L), transforming growth factor alpha (TGFα), tumour necrosis factor alpha (TNFα), tumour necrosis factor beta TNFβ and vascular endothelial growth factor A (VEGF-A). TGF-β 3-Plex Array Multi-Species (Eve Technologies, Calgary, Alberta, Canada) was also performed to analyse the tumour growth factor beta 1, 2 and 3 (TGF-β 1, 2 and 3).

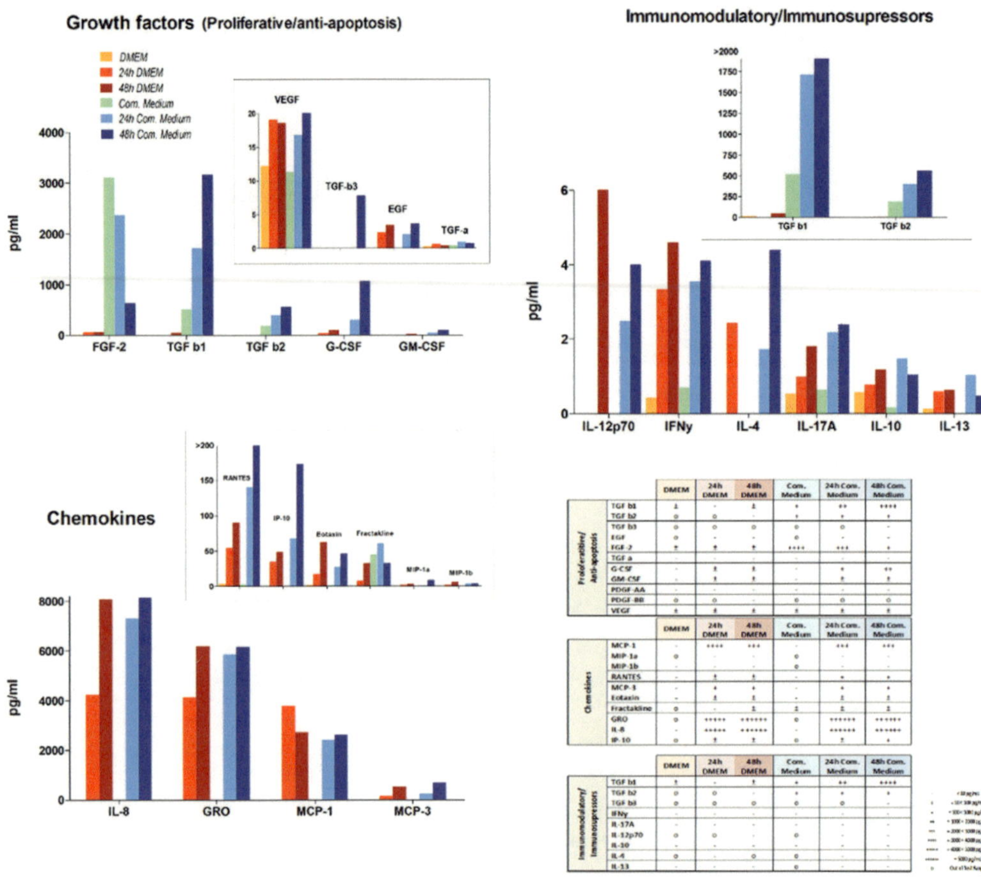

Figure 2. Proliferative and antiapoptotic growth factors, immunomodulatory, immunosupressive cytokines and chemokines concentrations in unconditioned (Com. Medium* and DMEM) and conditioned media (24/48 h Com. Medium and DMEM). (Multiplexing LASER Bead Analysis (Eve Technologies, Calgary, Alberta, Canada). *Commercial medium from PromoCell (LabClinics, Promocell, reference C-28010), adapted from [31].

The results from the Human Primary Cytokine Array/Chemokine Array 41-Plex Panel together with the TGF-β 3-Plex Array allowed the comprehensive analysis of the cytokines, chemokines and growth factors included in conditioned media in comparison with unconditioned culture media [43]. It was confirmed that the concentration of important proliferative and anti-apoptotic factors increases the process of cell conditioning. This was particularly evident for TGF-β1, EGF, G-CSF, GM-CSF, PDGF-AA and VEGF. Some of these factors like EGF were even

http://dx.doi.org/10.5772/66101

absent in unconditioned media. It was also confirmed that the content of factors such as TGF-β1 and G-CSF depends on the duration of the conditioning process (**Figure 2**).

4.2. Proteomic analysis of hUCBP

Following an identical analytical protocol hUCBP depicted impressively high detection levels for TGF-β family members 1–3. Fair concentrations of other factors like VEGF, PDGF-AA, PDGF-BB and EGF were also observed. Other immunossupressive/immunomodulatory factors can also be found at interesting levels in hUCBP (**Figure 3**).

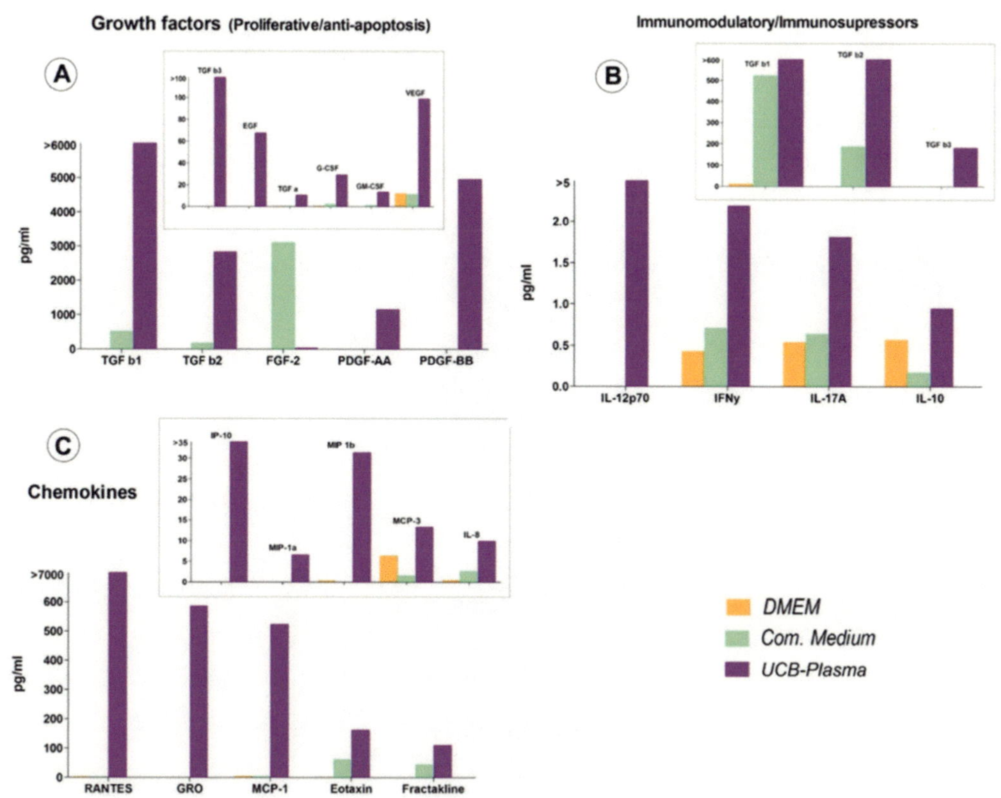

Figure 3. Proliferative and anti-apoptotic growth factors (A), immunomodulatory, immunosuppressive cytokines (B) and chemokines (C) concentrations in UCB Plasma and unconditioned media (Com. Medium* and DMEM). (Multiplexing LASER Bead Analysis (Eve Technologies, Calgary, Alberta, Canada). *Commercial medium from PromoCell (LabClinics, Promocell, C-28010), as in [31].

5. Metabolomics techniques

Metabolomics is a powerful approach for multicomponent analysis of biological samples that allow the comprehensive and simultaneous systematic profiling of multiple metabolite concentrations and their cellular and systemic fluctuations in response to xenobiotic exposure, environment, stimuli and genetic modulations. Metabolomics is important and well-established technique that encompasses the multicomponent analysis of metabolites in biological fluids, tissue and cell extracts using advanced analytical techniques [21, 43–46].

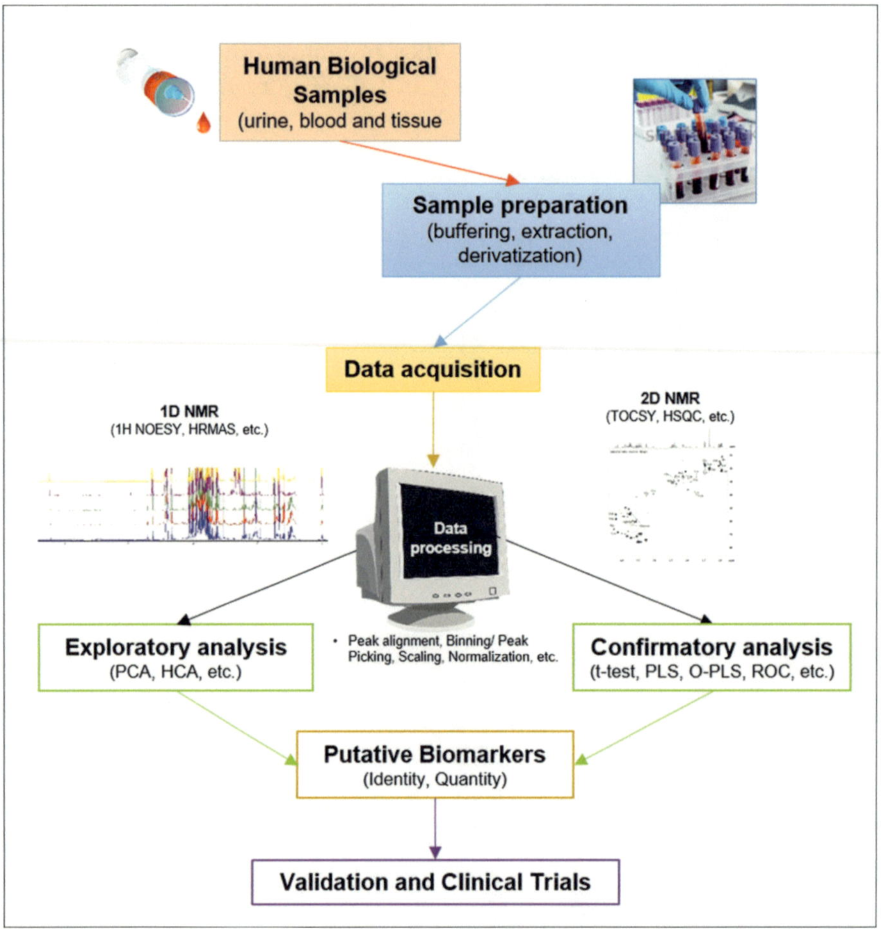

Figure 4. Schematic representation of the steps followed in a metabolomic analysis of conditioned medium obtained from MSCs' *in vitro* expansion.

An analysis of the metabolite spectrum in a biological system provides a detailed information and specific view into cellular metabolic processes under normal and altered conditions. The study of the entire multicomponent metabolic profile of biological systems is a challenging task and relies on the use of reproducible analytical technologies that provide sufficiently high sensitivity, high resolution and wide dynamic range. Metabolomics is mainly based on NMR spectroscopy (NMR) and mass spectrometry (MS), with chromatographic separation usually involved in the latter [43–46]. Both spectroscopic techniques could give valuable information on the structure and quantitative distribution of metabolites in biological systems.

Metabolomics studies usually result in large number of complex multivariate data sets that are typically analysed using chemometric and bioinformatics methods for data interpretation and visualization. These studies have been extensively expanded after implementation of pattern recognition (PR), expert systems and related bio-informatics tools to interpret and classify complex metabolic data sets. Well-established methodologies and protocols for metabolic analysis of biological samples have been extensively developed [21, 43–48].

The metabolomics analysis of biological samples involves several basic steps (**Figures 1** and **4**): sample collection; sample manipulation and storage; sample preparation for metabolic analysis; application of appropriate analytical method(s); data acquisition; data analysis (metabolites identification and quantification) and pattern recognition of spectral data (data interpretation and visualisation).

The development of advanced analytical techniques and instrumentation in the last decades has enabled a comprehensive metabolic analysis of diverse and complex biological samples, such as tissues, cells, tissues and cellular extracts, bio-fluids, etc. Certain advanced analytical techniques such as NMR, MS, LC, LC-MS and bioinformatics exist for metabolic analysis of biological systems, which can be applied in MSC secretomics [21, 43–48].

Studies of the composition of MSCs' secretome could contribute to the understanding of the mechanisms underlying the therapeutic effects of MSCs and assessment of the potential of MSCs' secretome as an alternative to MSCs' cell therapy.

5.1. Metabolic analysis of MSCs' secretome by NMR spectroscopy

NMR spectroscopy being a high reproducible and quantitative analytical technique, with minimal or no sample preparation and the sample can be recovered following the analysis, is one of the main tools for metabolic profiling, identification and quantification of known and unknown metabolites in biological samples. It offers unique opportunities for improving the structural and functional characterization of metabolome, which can be essential for the understanding of many biological processes and physiological changes in biological systems [43–48].

NMR spectroscopy as a quantitative non-destructive, non-invasive analytical technique can be used effectively to screen for metabolite profiles and to monitor structural and physiological changes in biological systems [21, 43, 44]. Nowadays, high-resolution NMR spectroscopy is one of the most powerful analytical techniques available for metabolic profiling of biological systems and could be suitable for metabolic analysis of MSC secretome. One of the advantages of NMR spectroscopy over other methods is that it can generate a large quantity of information concerning the metabolic composition of samples, making possible the simultaneous identification and quantification of structurally diverse metabolites, without the need for individual isolation or no special sample preparation [43, 44].

5.1.1. Samples collection and preparation

As for the proteomic analysis, the sample collection, storage and processing procedures are critical for conducting successful metabolomics studies. As previously studied, the collected samples of CM or for other biological fluids like umbilical cord blood plasma [31] should be centrifuged to remove cellular detritus and stored at –20 or even –80°C. The cellular systems used in the metabolomics and secretome profile studies should be previously characterised, including the flow cytometry or immunecitochemistry profile and karyotype.

At the end of the cell conditioning period, the conditioned medium is collected and centrifugation and/or filtration are usually applied to remove cell residues. In some cases, larger volumes of conditioned medium can be concentrated to succeed optimal experimental conditions. Upon collection and purification, the conditioned medium can be frozen and stored at –20°C, being later on thawed on the day of the NMR experiments.

The preparation of samples for NMR analysis can involve the addition of buffer to stabilize the pH, deuterated water (D_2O) as a magnetic field lock signal and reference compound to be used as a chemical shift and quantitation standard. The reference compound used in aqueous media is usually sodium trimethylsilyl-[2,2,3,3-d4]-propionate (TSP), which can be used as an internal or external standard. Absolute concentrations of metabolites can be obtained by the use of internal standard of known concentration added to the sample.

5.1.2. NMR techniques

The metabolomic study of bio-fluids is usually based of ^{1}H NMR spectroscopy. The NMR data for all samples need to be acquired and processed uniformly. The NMR experimental conditions and procedures are critical for conducting successful metabolomics studies and need to be uniform. The large interfering NMR signal arising from water in all bio-fluids can be eliminated by the use of standard NMR solvent suppression pulse sequences. ^{1}H NMR spectra using a 1D NOESY (noesygppr1d) pulse sequence (recycle delay-90°-t1-90°-tm- acquire) and a Carr-Purcell-Meiboom-Gill (CPMG, cpmgpr1d) pulse sequence (recycle delay-90°-[τ-180°-τ]n- acquire) can be acquired. The 1D NOESY experiment generates spectra with improved solvent peak (at 4.70 ppm) suppression. Relaxation edited ^{1}H NMR spectra with T2 (spin-spin relaxation time) filter using CPMG pulse sequence and suppression of water resonance were acquired to facilitate the identification of low molecular weight metabolites, reducing signals from high molecular weight species or systems in intermediate chemical exchange. The generated NMR data should be processed spectra were processed uniformly.

5.1.3. NMR data analysis

NMR spectra of biological systems are usually extremely complex because of the large number of components in the samples, resulting in spectra with complex line shapes and significant overlap of the resonance signals. The analysis of NMR spectra and identification of the metabolites in biological samples may involve a number of spectral techniques, information from databases of known metabolite spectra and NMR assignment software. Though very complex, the NMR spectra of biological samples (in particular ^{1}H NMR) allow direct assignment of resonance signals of some metabolites based on their chemical shifts, multiplicity and intensity and spectral analysis of appropriate NMR spectroscopic techniques such as two (2D) or multidimensional NMR techniques. 2D NMR spectroscopy can be useful to increase the signal dispersion in spectra with significant overlap of the resonances and reveal the connectivity between signals (^{1}H/1H Correlation spectroscopy (COSY); ^{1}H/^{1}H Total Correlation spectroscopy (TOCSY), ^{1}H/^{13}C or ^{1}H/^{15}N Hetero-correlation spectroscopy (HSQC, HMBC)) and/or to define the multiplicity and coupling pattern of resonances (2D J-resolved spectroscopy), thereby helping to identify metabolites in biological samples [21, 43, 44, 48].

5.1.4. Bioinformatics

In NMR metabolomics, the interpretation and classification of the complex metabolic data sets typically require the use of chemometric and pattern recognition (PR) techniques, and related bio-informatics tools. Pattern recognition techniques such as principal component analysis (PCA) allow for the analysis and classification of the large number of complex NMR spectroscopic data usually acquired for metabolic profiling of biological samples [44, 47, 48].

5.1.5. Applications of NMR–based metabolomics

NMR-metabolomics has been applied to monitor cellular and systemic metabolic fluctuations in response to xenobiotic exposure, environmental factors, physiological stimuli and genetic modifications [21, 43–48]. Many applications of NMR-metabolomics have been published, including the extensive study of physiological variation in experimental animals, age-related changes, dietary modulation, diurnal effects and phenotyping of mutant and transgenic animals [44, 47, 49].

NMR-based metabolomics has found many applications in metabolic analysis of a wide diversity of biological samples and has been recently used for characterization, monitoring and optimisation of stem cell culture medium preparation [50]. NMR spectroscopy was used for metabolomics analysis of human embryonic stem cell (hESCs) conditioning medium characterization. A number of metabolites were identified and quantified in conditioned media. Significantly, higher concentrations for certain metabolites (lactate, alanine, glutamine, glucose and formate) and lower for others (tryptophan, folate and niacinamide) were detected in the conditioned media compared to unconditioned media. Multivariate statistical analysis was applied to classify the data and assess the main metabolic changes in the process of cell conditioning. It was confirmed that NMR could be an accurate and valuable tool for monitoring, controlling and optimising hESC culture medium preparation. In this study, the detailed scrutiny of metabolites involved in a variety of biochemical pathways for potential tissue engineering applications has been facilitated [50].

In a more recent study by the application of proton NMR spectroscopy and implementation of appropriate one-dimensional (1D) and two-dimensional (2D) NMR techniques, it was confirmed that the metabolic composition of Wharton's jellies derived MSCs' conditioned media and changes in the metabolic profile of the culture media in the process of stem cell expansion and differentiation. These changes were found to be affected mainly by the type of the culture media and the duration of the conditioning time [31]. Typical ^{1}H NMR spectra of two different types of unconditioned and MSCs' conditioned media are presented in **Figure 5**.

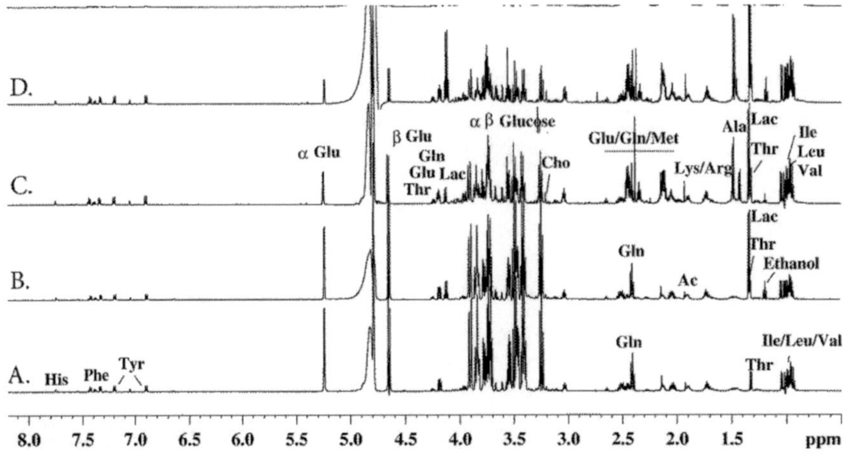

Figure 5. Typical 600 MHz ^{1}H NMR spectra recorded in H_2O:D_2O (9:1) of (A) Unconditioned medium I, (B) MSCs' conditioned medium I; (C) Unconditioned medium II, (D) MSCs' conditioned medium II. Characteristic signals of metabolites in the samples are marked.

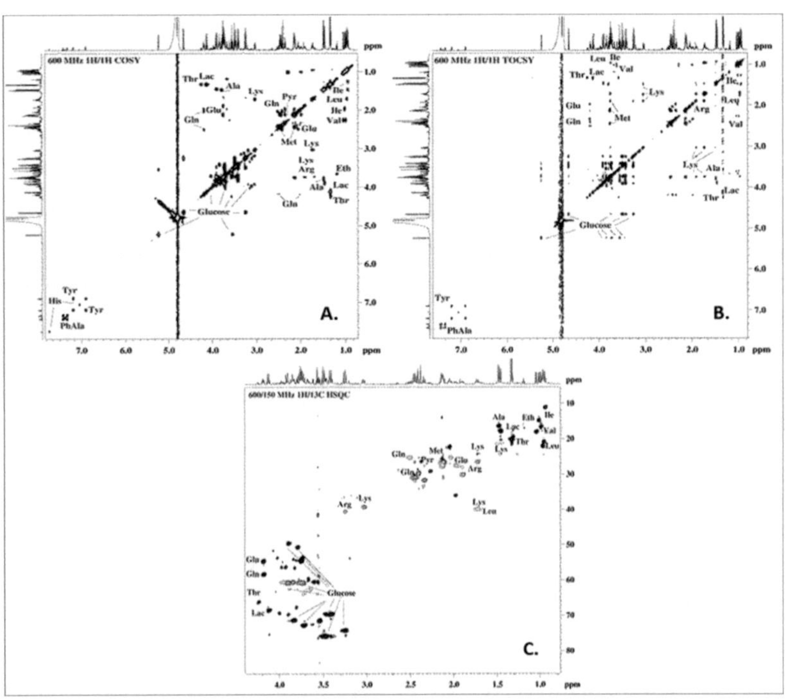

Figure 6. Representative 2D NMR experiments of MSCs' conditioned medium (conditioned medium I): (A) 1H/1H COSY spectrum, (B) 1H/1H TOCSY spectrum, (C) 1H/13C HSQC Spectrum.

The metabolomics study of different types culture media and the corresponding Wharton's jelly derived MSCs' conditioned media isolated at different time points was based on ^{1}H NMR spectroscopy. Sodium trimethylsilyl-[2,2,3,3-d4]-propionate (TSP) was used as an internal reference for the calibration and quantification of NMR spectra. ^{1}H NMR spectra with suppression of the water resonance signal using a 1D NOESY and CPMG pulse sequences were acquired. The NMR spectra were acquired and processed uniformly using TOPSPIN 3.2 (Bruker Biospin).

Additionally, to confirm the chemical shift assignment of ^{1}H NMR spectra and facilitate the identification of metabolites presented in both unconditioned and MSCs' conditioned media, the studied 2D $^{1}H/^{1}H$ COSY, $^{1}H/^{1}H$ TOCSY and $^{1}H/^{13}C$ HSQC spectra were obtained and analysed (**Figure 6**). The results have been compared to the data published in the literature [51].

Metabolites	Abbr.	Chemical shifts (ppm); multiplicity
Isoleucine	Ile	0.95(dd, δCH_3), 1.27/1.46(m, γCH_2), 1.98(m, βCH), 1.01(d, βCH_3), 3.67(d, αCH)
Leucine	Leu	0.95(d, δCH_3), 0.97(d, δ'CH_3), 1.69(m, γCH), 1.71(m, βCH_2), 3.73(d, αCH)
Valine	Val	0.98(d, γCH_3), 1.05(d, γ'CH_3), 2.27(m, βCH_2), 3.61(d, αCH)
Ethanol	Et	1.19 (t, CH_3), 3.57 (q, CH_2)
Threonine	Thr	1.33(d, γCH_3), 4.25(m, βCH_2), 3.56(d, αCH)
Lactate	Lac	1.34(d, βCH_3), 4.13(q, αCH)
Alanine	Ala	1.48(d, βCH_3), 3.78(q, αCH)
Lysine	Lys	3.04(t, εCH_2), 1.73(m, δCH_2), 1.47(m, γCH_2), 1.93(m, βCH_2), 3.77(t, αCH)
Arginine	Arg	3.23(t, δCH_2), 1.67(m, γCH_2), 1.89(m, βCH_2), 3.77(t, αCH)
Acetate	Ace	1.92(s, CH_3)
Glutamate	Glu	2.36(t, γCH_2), 2.08(m, βCH_2), 3.74 (dd, αCH)
Glutamine	Gln	2.44(t, γCH_2), 2.51/2.04(m, βCH_2), 3.76(dd, αCH)
Methionine	Meth	2.2(s, δCH_3), 2.45(m, γCH_2), 2.13(m, βCH_2), 3.75(t, αCH)
Choline	Cho	3.19(s, CH_3), 3.51(dd, CH_2), 4.06 (dd, CH_2),
Pyruvate	CH_3	2.39 (s)
α-Glucose	α-Glu	5.24(d,H1),3.54(dd,H2),3.72(t,H3),3.41(t,H4), 3.47(dd,H5),3.91/3,72(dd, H6)
β-Glucose	β-Glu	4.65(d,H1),3.24(t,H2),3.49(t,H3), 3.42(t,H4), 3.84(m,H5,6), 3.78(dd,H6)
Tyrosine	Tyr	3.05/3.18 (βCH_2), 3.93 (αCH), 7.19(d, H2,6), 6.90(d, H3,5)
Phenylalanine	PheAla	3.12/3.26 (βCH_2), 3.97 (αCH), 7.33(d, H2,6), 7.38(t,H4), 7.43(t, H3,5)
Histidine	His	3.12/3.22 (βCH_2), 3.97 (αCH), 7.75(s, H2), 7.05(s, H5)
Tryptophan	Try	3.29/3.47 (βCH_2), 4.05 (αCH), 7.30 (H2), 7.74(d, H4), 7.20(t, H5), 7.28(t, H6), 7.55(d, H7)
Nicotinamide	NA	7.58 (dd, CH), 8.24 (dd, CH), 8.70 (dd, CH), 8.94 (s, CH)
Formate	For	8.46 (CH)

Table 1. ^{1}H NMR chemical shifts and multiplicity of the main metabolites observed in the spectra of culture medium and MSCs' secretome.

From the analysis of the 1H and 2D ($^1H/^1H$ homonuclear and ^1H, ^{13}C heteronuclear correlation spectra) NMR spectra, certain number of metabolites have been identified. The main metabolites identified in the samples are listed in **Table 1** with an assignment of the corresponding resonance signals and their multiplicity in ^1H NMR spectra recorded.

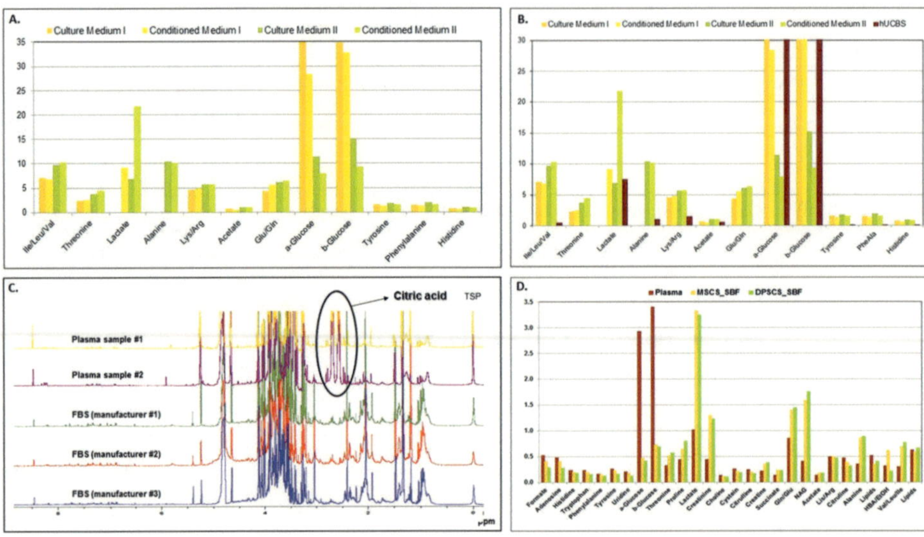

Figure 7. (A) Relative distribution (%) of the main metabolites detected in the culture media I and II and the corresponding MSCs' conditioned media I and II [31]; (B) Relative distribution (%) of the main metabolites detected in the culture media I and II and the corresponding MSCs' conditioned media I and II, and umbilical cord blood plasma (hUCBP) [31]; (C) Typical 600 MHz ^1H NMR spectra recorded in $H_2O:D_2O$ (9:1) of two plasma samples obtained from umbilical cord blood (hUCBP) and foetal bovine serum (FBS) from three different manufacturers; (D) The quantitative distribution (ppm) of metabolites determined from the integral intensity of characteristic signals in ^1H NMR spectra considering the number of protons contributing to the intensity of the signal in umbilical cord blood plasma (hUCBP) and foetal bovine serum (SBF) from two different fabricants used of *in vitro* culture of MSCs from the umbilical cord matrix (MSCs_SBF) and dental pulp mesenchymal stem cells (DPSCs_SBF).

The metabolite content in two different types of culture media (culture media I and II) and the corresponding Wharton's jelly derived MSCs' conditioned media collected at different time points was defined by quantitative NMR analysis, using the relative quantitative method [52]. The quantitative distribution of the NMR-visible metabolites was determined from the relative integral intensity of characteristic signals in ^1H NMR spectra of the samples referenced to the integral intensity of TSP (internal standard), considering the number of the contributing nuclei for that particular resonance. The relative quantitative distribution of the main metabolites detected in the ^1H NMR spectra of the two different types of culture media (culture medium I and culture medium II) and the corresponding Wharton's jelly derived MSCs' conditioned media collected after 24 h of conditioning is presented graphically in **Figure 7A**.

The results from the NMR metabolomics analysis reveal the presence of important amino acids, glucose and low molecular weight (lactate, acetate, pyruvate and formate) compounds in the basal culture media (types I and II) and the corresponding conditioned media.

Different contents of glucose, lactate and certain amino acids were detected in the two basal culture media (types I and II) and the corresponding conditioned media (**Figures 3** and **7A**). Gathering the results of unconditioned media (types I and II) and comparing them globally with the corresponding conditioned media, it is possible to attest not only to metabolite depletion through cell metabolisation (as the glucose example), but also to some metabolites (like lactate, isoleucine, leucine, valine, arginine, lysine, glutamate, formate or pyruvate) presented in higher concentration, reflecting the production and secretion of metabolites. Metabolic differences between samples collected at different time points (24 and 48 h) were also considered [31]. This is because conditioned media could be achieved at different stages after reaching minimal cell confluence.

5.2. Metabolic analysis of hUCBP by NMR spectroscopy

The metabolite profiles of the unconditioned and MSCs' conditioned media (types I and II) were further compared with those of human umbilical cord blood plasma (hUCBP) (**Figure 7B**) which has been considered as a promising culture medium supplement due to the fact that UCB is rich in soluble growth factors that support the growth, proliferation and differentiation of mesenchymal and hematopoietic stem cells [31, 33, 34].

The comparative study of the metabolic profiles of unconditioned media, MSCs' conditioned media and hUCBP reveals the presence of essential proteinogenic amino acids (alanine, arginine, isoleucine, histidine, leucine, lysine, phenylalanine, proline, tyrosine, valine), organic acids and derivatives (lactate, formate and creatine), glucose and nucleotides (adenosine, uridine and guanosine species) in the last. The results strongly suggest that hUCBP might be used as a culture medium supplement in the processes of cell isolation, expansion and cryopreservation and even as an alternative in the cellular therapy [31]. Moreover, hUCBP could be considered as an animal sera substituent that could solve important economic, ethical and scientific issues that have severe complications in the utilization of foetal bovine serum (FBS) for cell culture. In addition, hUCBP are attractive alternatives to FBS due to the worldwide increase in the number of cryopreserved UCB units in Public and Private Cord Blood Banks. However, the substitute medium supplement must possess the animal sera characteristics crucial for the cell expansion and preservation.

Recently, a comparative study of the metabolic composition of hUCBP and FBS, based on ^{1}H NMR spectroscopy and multivariate statistical analysis has been performed to evaluate the capability of hUCBP as a culture medium supplement and FBS substituent for *in vitro* proliferation and cryopreservation of MSCs. **Figure 7C** shows typical ^{1}H NMR spectra of hUCBP and FBS.

However, 1D and 2D NMR techniques have been used to identify and quantify the detectible metabolites in the samples. The quantitative distribution of metabolites presented in **Figure 7D** was determined from the integral intensity of characteristic signals in ^{1}H NMR spectra considering the number of protons contributing to the intensity of the signal.

The results show that the differences between the metabolic composition of hUCBP and FBS are due mainly to the different content of glucose, lactate, acetate, glutamate and alanine. The much higher content of glucose determined for hUCBP has been attributed to the UCB function to provide energetic support for the foetus during intrauterine development. From the analysis of the NMR spectra it was confirmed the presence of essential proteinogenic amino acids such as alanine, arginine, creatine, glutamine, glutamate, isoleucine, histidine, leucine, lysine, phenylalanine, proline, threonine, tyrosine and valine in hUBCP as well as in FBS. These proteinogenic amino acids are precursors to proteins and could play a crucial role in the growth, proliferation and differentiation of cells. Significantly, higher levels of alanine, glutamate, isoleucine, leucine and valine in the commercial FBS as compared to hUBCP are determined. This seems to be the main difference between the amino acids profiles of human UBC plasma and foetal bovine sera. The increased levels of alanine and glutamine could due to the use of supplements such as Glutamax® in the commercial FBS sera [50]. Similar or slightly higher contents of arginine, citrulline, creatine, histidine, lysine, phenylalanine and tyrosine, nucleotides and lipids were found in hUBCP as compared to the FBS samples studied.

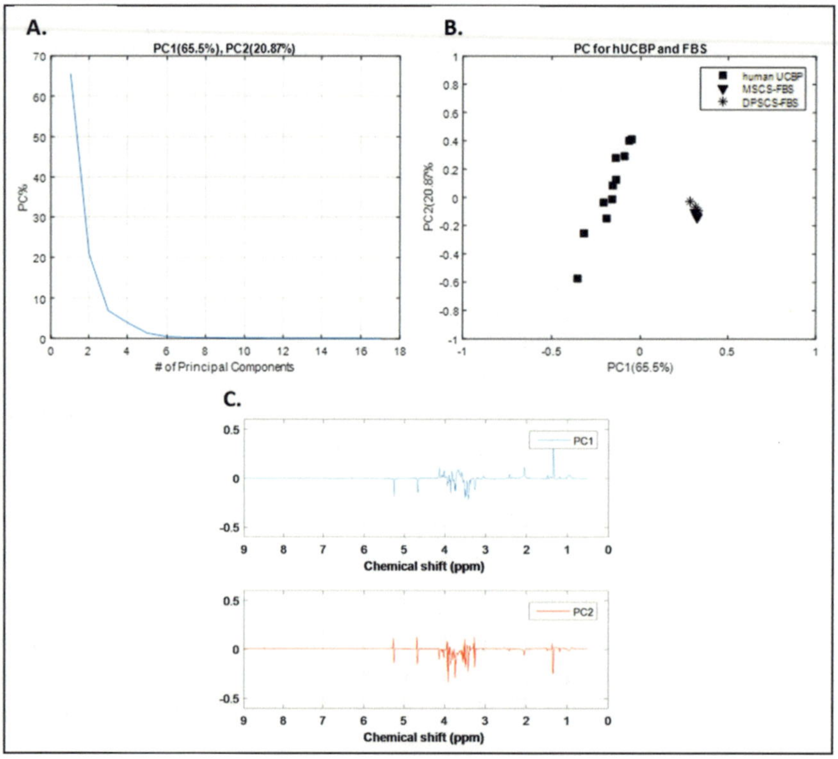

Figure 8. A and B—PCA statistics. PCs distribution of NMR data matrix (A) and Score plots (PC1 versus PC2) (B) and —PC1 and PC2 loading plots (C).

Principal component analysis has been used to interpret and classify the NMR spectral data obtained for a certain number of hUCBP (samples of 11 donors) and FBS samples (MSCs_FBS (3 samples of MSCs FBS and three samples of DPSCS FBS) to evaluate the main factors contributing to the discrimination between the two groups of samples.

http://dx.doi.org/10.5772/66101

The contributions of the principal components (PCs) of the NMR spectral data are depicted in **Figure 8A** and **B**. The first two PCs accumulate more than 85 % of data variance. Therefore statistical analysis based on PC1 and PC2 is legitimate.

The 2D (PC1 versus PC2) PCA score plot is depicted in **Figure 8B**. The results illustrate the close similarity of MSCs_FBS and DPSCS_FBS data sets (a well-defined cluster with overlapping samples) and reveal their separation from the human plasma data (a scattered cluster).

The corresponding loading plots in **Figure 8C** provide evidence that the glucose, lactate and acetates are the metabolites contributing to the divergence between the human plasma and FBS data sets. The results also reveal the close content of essential proteinogenic amino acids in the human plasma and foetal bovine sera. Loading plots around zero line suggest close metabolic profiles in the spectral region from 9.0 to 5.3 ppm of the three groups.

6. Conclusions

In brief conclusion, proteomic and metabolomics studies have demonstrated the richness of MSC' conditioned medium in terms of growth factors with *in situ* action in disease and regeneration processes, as well as in other metabolites that interact with the environment surrounding the mesenchymal cellular populations. Hence, as knowledge is gained in its composition and potential, thought these omics techniques and other assays, their potential on the application for the treatment of number of diseases increases alike.

As for the hUCBP, given its similarity and even superiority (in supplementary proteins and other contents) to promote and sustain MSCs' expansion, it might become a standardized human derived supplement for cellular populations aiming at the production of secretome or direct application in effective clinical treatments.

Acknowledgements

This research was supported by Programa Operacional Regional do Norte (ON.2 - O Novo Norte), QREN, FEDER with the project "iBone Therapies: Terapias inovadoras para a regeneração óssea", ref. NORTE-01-0247-FEDER-003262, and by the program COMPETE - Programa Operacional Factores de Competitividade, Projects PEst-OE/AGR/UI0211/2011, PEst-C/EME/UI0285/2013, and Pest-C/EQB/LA0006/2013 funding from FCT. This research was also supported by Programa Operacional Competitividade e Internacionalização (P2020), Fundos Europeus Estruturais e de Investimento (FEEI) and FCT with the project "BioMate - A novel bio-manufacturing system to produce bioactive scaffolds for tissue engineering" with reference PTDC/EMS-SIS/7032/2014 and by COMPETE 2020, from ANI - Projectos ID…T Empresas em Copromoção, Programas Operacionais POCI, by the project "insitu.Biomas - Reinvent biomanufacturing systems by using an usability approach for *in situ* clinic temporary implants fabrication" with the reference POCI-01-0247-FEDER-017771. Ana Rita Caseiro (SFRH/BD/

101174/2014) acknowledges FCT, for financial support. This work also received financial under the framework of QREN through Project NORTE-07-0124-FEDER-000066. The Bruker Avance III 600 HD spectrometer was purchased under the framework of QREN, through Project NORTE-07-0162-FEDER-000048, and is part of the Portuguese NMR Network created with support of FCT through Contract REDE/1517/RMN/2005, with funds from POCI 2010 (FEDER).

Author details

Galya Ivanova[1], Tiago Pereira[2,3], Ana Rita Caseiro[2,3,4], Petia Georgieva[5] and Ana Colette Maurício[1,4*]

*Address all correspondence to: ana.colette@hotmail.com

1 REQUIMTE-UCIBIO, Department of Chemistry and Biochemistry, Faculty of Sciences, Porto University, Porto, Portugal

2 Department of Veterinary Clinics, Abel Salazar Institute of Biomedical Sciences (ICBAS), Porto University (UP), Porto, Portugal

3 Animal Science Studies Centre (CECA), Agro-Alimentary Technologies and Sciences Institute, (ICETA), Porto, Portugal

4 CEMUC, Department of Materials and Metallurgic Engineering, Faculty of Engineering, Porto University, Porto, Portugal

5 Department of Electronics Telecommunications and Informatics, IEETA, University of Aveiro, Portugal

References

[1] Bianco P, Robey PG, Simmons PJ. Mesenchymal stem cells: revisiting history, concepts, and assays. Cell Stem Cell. 2008;2(4):313–9.

[2] Dominici M, Le Blanc K, Mueller I, Slaper-Cortenbach I, Marini F, Krause D, et al. Minimal criteria for defining multipotent mesenchymal stromal cells. The International Society for Cellular Therapy position statement. Cytotherapy. 2006;8(4):315–7.

[3] Barry FP, Murphy JM. Mesenchymal stem cells: clinical applications and biological characterization. The International Journal of Biochemistry & Cell Biology. 2004;36(4): 568–84.

[4] Pittenger MF, Mackay AM, Beck SC, Jaiswal RK, Douglas R, Mosca JD, et al. Multilineage potential of adult human mesenchymal stem cells. Science. 1999;284(5411):143–7.

[5] Mendicino M, Bailey AM, Wonnacott K, Puri RK, Bauer SR. MSC-based product characterization for clinical trials: an FDA perspective. Cell Stem Cell. 2014;14(2): 141–5.

[6] Giordano A, Galderisi U, Marino IR. From the laboratory bench to the patient's bedside: an update on clinical trials with mesenchymal stem cells. Journal of Cellular Physiology. 2007;211(1):27–35.

[7] Lai RC, Yeo RWY, Tan SS, Zhang B, Yin Y, Sze NSK, et al. Mesenchymal stem cell exosomes: the future MSC-based therapy? Mesenchymal Stem Cell Therapy: Springer; 2013. pp. 39–61.

[8] Friedenstein AJ, Petrakova KV, Kurolesova AI, Frolova GP. Heterotopic of bone marrow. Analysis of precursor cells for osteogenic and hematopoietic tissues. Transplantation. 1968;6(2):230–47.

[9] Friedenstein AJ, Chailakhyan RK, Latsinik NV, Panasyuk AF, Keilissb IV. Stromal cells responsible for transferring microenvironment of hematopoietic tissues–cloning in vitro and retransplantation in vivo. Transplantation. 1974;17(4):331–40.

[10] da Silva Meirelles L, Chagastelles PC, Nardi NB. Mesenchymal stem cells reside in virtually all post-natal organs and tissues. Journal of Cell Science. 2006;119(11): 2204–13.

[11] Carvalho MM, Teixeira FG, Reis RL, Sousa N, Salgado AJ. Mesenchymal stem cells in the umbilical cord: phenotypic characterization, secretome and applications in central nervous system regenerative medicine. Current Stem Cell Research & Therapy. 2011;6(3):221–8.

[12] Patel AN, Vargas V, Revello P, Bull DA. Mesenchymal stem cell population isolated from the subepithelial layer of umbilical cord tissue. Cell Transplant. 2013;22(3):513–9.

[13] Hass R, Kasper C, Bohm S, Jacobs R. Different populations and sources of human mesenchymal stem cells (MSC): a comparison of adult and neonatal tissue-derived MSC. Cell Communication and Signaling. 2011;9:12.

[14] Brunstein CG, Setubal DC, Wagner JE. Expanding the role of umbilical cord blood transplantation. British Journal of Haematology. 2007;137(1):20–35.

[15] Phinney DG, Prockop DJ. Concise review: mesenchymal stem/multipotent stromal cells: the state of transdifferentiation and modes of tissue repair—current views. Stem Cells. 2007;25(11):2896–902.

[16] Caplan AI, Dennis JE. Mesenchymal stem cells as trophic mediators. Journal of Cellular Biochemistry. 2006;98(5):1076–84.

[17] Zimmerlin L, Park TS, Zambidis ET, Donnenberg VS, Donnenberg AD. Mesenchymal stem cell secretome and regenerative therapy after cancer. Biochimie. 2013;95(12): 2235–45.

[18] Mustafa SA, Hoheisel JD, Alhamdani MS. Secretome profiling with antibody microarrays. Molecular Biosystems. 2011;7(6):1795–801.

[19] Skalnikova H, Motlik J, Gadher SJ, Kovarova H. Mapping of the secretome of primary isolates of mammalian cells, stem cells and derived cell lines. Proteomics. 2011;11(4): 691–708.

[20] Skalnikova HK. Proteomic techniques for characterisation of mesenchymal stem cell secretome. Biochimie. 2013;95(12):2196–211.

[21] Lindon JC, Nicholson JK. Analytical technologies for metabonomics and metabolomics, and multi-omic information recovery. Trac–Trends in Analytical Chemistry. 2008;27(3): 194–204.

[22] Malmstrom J, Lee H, Aebersold R. Advances in proteomic workflows for systems biology. Current Opinion in Biotechnology. 2007;18(4):378–84.

[23] Brown KJ, Formolo CA, Seol H, Marathi RL, Duguez S, An E, et al. Advances in the proteomic investigation of the cell secretome. Expert Review of Proteomics. 2012;9(3): 337–45.

[24] Panfoli I, Calzia D, Santucci L, Ravera S, Bruschi M, Candiano G. A blue dive: from 'blue fingers' to 'blue silver'. A comparative overview of staining methods for in-gel proteomics. Expert Review of Proteomics. 2012;9(6):627–34.

[25] Haynesworth SE, Baber MA, Caplan AI. Cytokine expression by human marrow-derived mesenchymal progenitor cells in vitro: effects of dexamethasone and IL-1a. Journal of Cellular Physiology. 1996;166(3):585–92.

[26] Lai RC, Arslan F, Tan SS, Tan B, Choo A, Lee MM, et al. Derivation and characterization of human fetal MSCs: an alternative cell source for large-scale production of cardioprotective microparticles. Journal of Molecular and Cellular Cardiology. 2010;48(6): 1215–24.

[27] Timmers L, Lim SK, Hoefer IE, Arslan F, Lai RC, van Oorschot AAM, et al. Human mesenchymal stem cell-conditioned medium improves cardiac function following myocardial infarction. Stem Cell Research. 2011;6(3):206–14.

[28] Brion C, Miller SG, Moore HPH. Regulated and constitutive secretion—differential-effects of protein-synthesis arrest on transport of glycosaminoglycan chains to the 2 secretory pathways. Journal of Biological Chemistry. 1992;267(3):1477–83.

[29] Lee MJ, Kim J, Kim MY, Bae YS, Ryu SH, Lee TG, et al. Proteomic analysis of tumor necrosis factor-alpha-induced secretome of human adipose tissue-derived mesenchymal stem cells. Journal of Proteome Research. 2010;9(4):1754–62.

[30] Kilroy GE, Foster SJ, Wu X, Ruiz J, Sherwood S, Heifetz A, et al. Cytokine profile of human adipose-derived stem cells: expression of angiogenic, hematopoietic, and pro-inflammatory factors. Journal of Cellular Physiology. 2007;212(3):702–9.

[31] Pereira T, Ivanova G, Caseiro AR, Barbosa P, Bartolo PJ, Santos JD, et al. MSCs conditioned media and umbilical cord blood plasma metabolomics and composition. PLoS One. 2014;9(11):e113769.

[32] Tekkatte C, Gunasingh GP, Cherian KM, Sankaranarayanan K. "Humanized" stem cell culture techniques: the animal serum controversy. Stem Cells International. 2011;2011.

[33] Huang L, Critser PJ, Grimes BR, Yoder MC. Human umbilical cord blood plasma can replace fetal bovine serum for in vitro expansion of functional human endothelial colony-forming cells. Cytotherapy. 2011;13(6):712–21.

[34] Bieback K, Hecker A, Kocaomer A, Lannert H, Schallmoser K, Strunk D, et al. Human alternatives to fetal bovine serum for the expansion of mesenchymal stromal cells from bone marrow. Stem Cells. 2009;27(9):2331–41.

[35] Potian JA, Aviv H, Ponzio NM, Harrison JS, Rameshwar P. Veto-like activity of mesenchymal stem cells: functional discrimination between cellular responses to alloantigens and recall antigens. Journal of Immunology. 2003;171(7):3426–34.

[36] Kim HS, Choi DY, Yun SJ, Choi SM, Kang JW, Jung JW, et al. Proteomic analysis of microvesicles derived from human mesenchymal stem cells. Journal of Proteome Research. 2012;11(2):839–49.

[37] Bernay B, Gaillard MC, Guryca V, Emadali A, Kuhn L, Bertrand A, et al. Discovering new bioactive neuropeptides in the striatum secretome using in vivo microdialysis and versatile proteomics. Molecular & Cellular Proteomics. 2009;8(5):946–58.

[38] Sze SK, de Kleijn DPV, Lai RC, Tan EKW, Zhao H, Yeo KS, et al. Elucidating the secretion proteome of human embryonic stem cell-derived mesenchymal stem cells. Molecular & Cellular Proteomics. 2007;6(10):1680–9.

[39] Gilar M, Olivova P, Daly AE, Gebler JC. Two-dimensional separation of peptides using RP-RP-HPLC system with different pH in first and second separation dimensions. Journal of Separation Science 2005;28(14):1694–703.

[40] Yang F, Shen YF, Camp DG, Smith RD. High-pH reversed-phase chromatography with fraction concatenation for 2D proteomic analysis. Expert Review of Proteomics. 2012;9(2):129–34.

[41] Chen XL, Cushman SW, Pannell LK, Hess S. Quantitative proteomic analysis of the secretory proteins from rat adipose cells using a 2D liquid chromatography-MS/MS approach. Journal of Proteome Research. 2005;4(2):570–7.

[42] Zhang X, Fang AQ, Riley CP, Wang M, Regnier FE, Buck C. Multi-dimensional liquid chromatography in proteomics–a review. Analytica Chimica Acta. 2010;664(2):101–13.

[43] Beckonert O, Keun HC, Ebbels TM, Bundy J, Holmes E, Lindon JC, et al. Metabolic profiling, metabolomic and metabonomic procedures for NMR spectroscopy of urine, plasma, serum and tissue extracts. Nature Protocols. 2007;2(11):2692–703.

[44] Zhang S, Nagana Gowda GA, Ye T, Raftery D. Advances in NMR-based biofluid analysis and metabolite profiling. Analyst. 2010;135(7):1490–8.

[45] Madhu B, Narita M, Jauhiainen A, Menon S, Stubbs M, Tavare S, et al. Metabolomic changes during cellular transformation monitored by metabolite-metabolite correlation analysis and correlated with gene expression. Metabolomics. 2015;11(6):1848–63.

[46] Beger RD. A review of applications of metabolomics in cancer. Metabolites. 2013;3(3): 552–74.

[47] Lindon JC, Holmes E, Nicholson JK. Pattern recognition methods and applications in biomedical magnetic resonance. Progress in Nuclear Magnetic Resonance Spectroscopy. 2001;39(1):1–40.

[48] Dona AC, Jimenez B, Schafer H, Humpfer E, Spraul M, Lewis MR, et al. Precision High-throughput proton NMR spectroscopy of human urine, serum, and plasma for large-scale metabolic phenotyping. Analytical Chemistry. 2014;86(19):9887–94.

[49] Abelev B, Adam J, Adamova D, Adare AM, Aggarwal MM, Aglieri Rinella G, et al. D meson elliptic flow in noncentral Pb-Pb collisions at sqrt[sNN] = 2.76 Tev. Physical Review Letters. 2013;111(10):102301.

[50] MacIntyre DA, Melguizo Sanchís D, Jiménez B, Moreno R, Stojkovic M, Pineda-Lucena A. Characterisation of human embryonic stem cells conditioning media by 1H-nuclear magnetic resonance spectroscopy. PLoS One. 2011;6(2):e16732.

[51] Nicholson JK, Foxall PJ, Spraul M, Farrant RD, Lindon JC. 750 MHz 1H and 1H-13C NMR spectroscopy of human blood plasma. Analytical Chemistry. 1995;67(5):793–811.

[52] Malz F, Jancke H. Validation of quantitative NMR. Journal of PHARMACeutical and Biomedical Analysis. 2005;38(5):813–23.

Chapter 4

Processing and Visualization of Metabolomics Data Using R

Stephen C. Grace and Dane A. Hudson

Additional information is available at the end of the chapter

http://dx.doi.org/10.5772/65405

Abstract

Metabolomics provides a wealth of information about the biochemical status of cells, tissues, and other biological systems. However, for many researchers, processing the large quantities of data generated in typical metabolomics experiments poses a formidable challenge. Robust computational tools are required for all data processing steps, from handling raw data to high level statistical analysis and interpretation. This chapter describes several established methods for processing and analyzing metabolomics data within the R statistical programming environment. The focus is on processing LCMS data but the methods can be applied virtually to any analytical platform. We provide a step-by-step workflow to demonstrate how to integrate, analyze, and visualize LCMS-based metabolomics data using computational tools available in R. These concepts and methods will allow specialists and nonspecialists alike to develop and evaluate their own data more critically.

Keywords: liquid chromatography mass spectrometry (LCMS), R, data processing, multivariate analysis, data visualization

1. Introduction

Metabolomics is a rapidly growing discipline focusing on the global study of small molecule metabolites in biological systems. Through the characterization of metabolite dynamics, interactions, and responses to genetic or environmental perturbations, metabolomics can provide a comprehensive picture of both baseline physiology and global biochemical responses to genetic, abiotic, and biotic factors [1].

As the diversity in abundance and chemical properties of metabolites varies greatly in organisms, a range of analytical techniques must be utilized to survey the entire metabolome. A number of methods have been developed for the extraction, detection, identification, and quantification of the metabolome [2]. Mass spectrometry coupled with gas chromatography (GC-MS) or liquid chromatography (LC-MS) are the most common analytical platforms, although capillary electrophoresis mass spectrometry (CE-MS) and nuclear magnetic resonance (NMR) are also widely used in metabolomics research [3–6].

Since metabolomics experiments typically produce large amounts of data, sophisticated bioinformatic tools are needed for efficient and high-throughput data processing to remove systematic bias and to explore biologically significant findings. Both multivariate statistical analysis and data visualization play a critical role in extracting relevant information and interpreting the results of metabolomics experiments.

The data generated in a metabolomics experiment generally can be represented as a matrix of intensity values containing N observations (samples) of K variables (peaks, bins, etc.). Additional information, such as experimental group, genotype, time point, gender, etc., is also required for some statistical procedures. For multivariate analysis, very few mathematical constraints are placed on the values contained in the data matrix. Therefore, a common set of statistical tools can be used to analyze metabolomics data of almost any type. However, as discussed below, multiple preprocessing steps are often necessary to yield interpretable results [7, 8].

The focus of this chapter is on describing methods for processing and visualizing metabolomics data obtained by liquid chromatography mass spectrometry (LCMS). LCMS is the most widely used method in metabolomics research due to its dynamic range, coverage, ease of sample preparation, and high information content [3–5]. We present a standard workflow for handling LCMS data, from raw data processing to downstream statistical analysis using open source tools available within the R software environment.

1.1. What is R?

R is a software environment for statistical computing, data analysis, and graphics, which has become an essential tool in all areas of bioinformatics research. A major advantage of R over commercial software is that it is open source and free to all users. The base distribution of R and a large number of user contributed packages are available under the terms of the Free Software Foundation's GNU General Public License in source code form. There are versions of R for Unix, Windows, and Macintosh at the official CRAN website (http://cran.r-project.org/).

In addition to being a popular language for performing high level statistics, R has a wide array of graphical tools that make it an ideal environment for exploratory data analysis and generating publication quality figures. All work is done using the command line-based text functions with user-defined scripts. Although R can be challenging for new users, it is quite flexible once the basic commands, functions, and data structures have been learned. A detailed description of every function with examples can be obtained by typing *help* followed by the name of the function, i.e., *help(plot)*. In addition, there are ample online resources to help users learn the basics of R as well as solve a wide range of common data analysis problems [9–11].

R has a powerful set of functions for creating graphics, from fairly simple graphs using base graphics commands to highly sophisticated graphs using the one of several advanced graphics packages [12]. The focus of this chapter is on using the ggplot2 package, a high-level platform for creating graphics that is especially powerful for working with high-dimensional data [13]. The basic idea in ggplot2 is to build graphs by adding successive layers that include visual representations as well as statistical summaries of the data [14]. A layer is defined as an R data frame or matrix, a specification mapping columns of that frame into aesthetic properties, a statistical approach to summarize the rows of that frame, a geometric object to visually represent that summary, and an optional position adjustment to move overlapping geometric objects. This approach allows great flexibility in producing highly customizable graphs by combining layers that describe single or multiple data objects. We will demonstrate these ideas throughout this chapter.

1.2. Open source tools for metabolomics

A number of free software tools are available for processing, visualization, and statistical analysis of metabolomics data. Some of the more popular platforms are presented in **Table 1**.

XCMS is a powerful R-based software for LCMS data processing. As with any R-based package, it is command line driven and requires some background knowledge of the R programming language. XCMS uses nonlinear retention time correction, matched filtration, peak detection, and peak matching to extract relevant information from raw LCMS data [15]. Peak detection parameters can be optimized to process the raw data in an appropriate and efficient manner. As shown in the following sections, XCMS can be combined with base R functions and additional R packages to provide a complete solution to LCMS data processing needs. Statistical analysis and data visualization can all be incorporated into the scripts to quickly process the large amounts of data from start to finish.

XCMS online is a web-based version of XCMS that provides many of the advantages of the traditional R package without the use of a command line-based environment [16]. It allows limited control over processing parameters and gives interactive graphs of univariate and multivariate analyses [17].

MetaboAnalyst is a popular web-based resource that provides an easy to use, comprehensive interface for metabolomics data analysis [18]. It includes a variety of data preprocessing and statistical tools for univariate and multivariate analysis and generates high resolution, interactive graphics. Depending on the type of data being analyzed, it can also be used for biomarker analysis, enrichment analysis, pathway analysis, and more [19].

Haystack is a web server-based processing tool that uses mass bins to filter and extract information from raw LCMS data [20]. Haystack also provides graphical tools to visualize raw and processed data and incorporates some exploratory statistical analysis tools. Because extracted features are based on mass bins, missing values due to redundant or missing peaks are absent from the processed data. Processed data files are downloadable in .csv format, which can be imported into analysis software of the user's preference.

Platform	Description	Advantages	Disadvantages	Reference
XCMS	R-based platform for raw LCMS data processing and visualization	-Adjustable parameters -Streamlined workflow	-Requires knowledge of R language -Command line based	[15]
XCMS online	Web-based graphical user interface version of XCMS	-Cloud storage and sharing -Relatively easy to use	-Not as customizable as the R version	[16, 17]
MetaboAnalyst	Online statistical analysis	-Easy to use -Wide variety of statistical tests available -Interactive plots	-Relies on pre-processed data -Limited options for customizing graphics	[18, 19]
Haystack	Raw data processing and visualization using mass bins	-Unbiased -No zero values -Not dependent on quality of chromatography	-Graphics not customizable -Does not take into account peak retention time	[20]
MZmine 2	Raw data processing and visualization	-Java based -User friendly -Project batching	-Limited options for customizing graphics -Numerous options can be overwhelming	[21]
MET-IDEA	Raw data processing for GCMS and LCMS data	-Works well with very large data sets -Optional manual integration	-Aimed more for GCMS data -Low-quality graphics	[22]

Table 1. Open source and web-based platforms for metabolomics data analysis.

MZmine 2 is a Java-based platform that allows for flexible MS data processing through a user-friendly graphical interface and customizable parameters for data processing and visualization [21].

Metabolomics Ion-Based Data Extraction Algorithm (MET-IDEA) is a large-scale metabolomics data processing program generally used for GCMS data but can also be used for LCMS data. It performs peak alignment, annotation, and integration of hyphenated mass spectrometry data and allows visualization of integrated peaks along with their accompanying mass spectra [22].

1.3. Data processing for metabolomics

Robust computational tools are essential to analyze and interpret metabolomics data. The first step in data processing, especially in untargeted metabolomics, is to convert the raw data into a numerical format that can be used for downstream statistical analysis. For LCMS data, this involves multiple steps, including filtering, feature detection, alignment, and normalization [23, 24]. Filtering methods aim to remove effects like measurement or baseline noise. Feature detection is used to identify measured ions from the raw signal. Alignment methods cluster measurements from across different samples and normalization removes unwanted systematic variation between samples.

Once the data have been converted to a numerical matrix, statistical tools are used to reveal patterns in the data, determine class membership, and identify relevant biological features. Univariate methods are often used as a first step to obtain a rough ranking of potentially important features before applying more sophisticated statistical techniques [25]. Familiar examples include fold change differences, *t*-tests, and volcano plots.

Multivariate methods treat multiple, often correlated, variables simultaneously, and attempt to model relationships between variables and observations [25–27]. Well known examples include principal component analysis (PCA) and partial least squares (PLS). PCA is an unsupervised method meaning that only the data matrix itself is used to model the data. Since class membership is not considered, PCA provides an unbiased summary of the data structure. For exploratory studies where differences between experimental groups may be unknown or unpredictable, it is appropriate to apply PCA as a first step to reveal patterns in the data and relationships between groups.

A shortcoming of PCA is that it can only reveal group differences when within-group variation is sufficiently less than between-group variation [26]. To overcome this problem, supervised forms of discriminant analysis such as PLS that rely on class membership are also routinely applied in metabolomics studies [28]. The primary goal of these methods is to identify class differences from a multivariate data set and to identify biologically important features that account for these differences. Often, the results of a PCA are used to formulate a hypothesis that PLS or other supervised methods can test or verify in more detail.

In the following sections, we will provide an overview of the data processing steps used in a typical metabolomics experiment. We do not aim to provide a thorough description of the statistical methods but rather introduce the basic concepts behind these methods and demonstrate how to perform them in the R computing environment.

2. Experimental methods

To illustrate the concepts and methods presented in this chapter, we produced a data set from two tomato varieties harvested at two developmental stages using untargeted LCMS. The varieties used were "Manapal" and its nearly isogenic counterpart, the *h*igh *p*igment-2 *dark green* (hp-2dg) mutant. Hp-2dg plants have a mutation in the tomato homolog of the DEETIOLATED-1 gene involved in light-mediated signal transduction and plant photomorphogenesis [29]. These plants display a number of interesting traits, including shorter stature, slower growth rates, darker foliage, and elevated levels of certain metabolites such as flavonoids and carotenoids [30–32]. Since many of these phytochemicals are important for human health, there is great interest in understanding the molecular mechanisms that underlie the altered phenotype of the hp-2dg mutation.

Fully expanded fruits were harvested at green and red stages, lyophilized, and stored at −80°C until analysis. Samples from 10 individual fruits were extracted in 80% methanol and analyzed using an Agilent 1100 HPLC/MSD-VI Ion Trap mass spectrometer with an electrospray ionization (ESI) source. Chromatograms were saved in netCDF format using the instrument software.

The data were analyzed using R as described in the following sections. A summary of the data processing workflow is presented in **Figure 1**. The full source code for all procedures is available online at https://github.com/ualr-Rgroup/metabolomics-in-r.

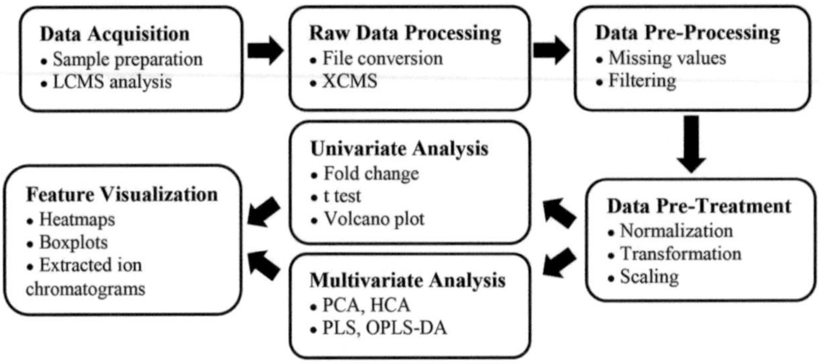

Figure 1. Summary of metabolomic data processing workflow.

2.1. Data processing with XCMS

While web-based tools such as MetaboAnalyst and XCMS online are inarguably convenient, learning data analysis procedures in R gives researchers much greater flexibility not only in processing and analyzing their data but also in creating high-quality custom graphics. Here, we demonstrate how to perform raw data processing in R using the XCMS package. XCMS is a powerful and flexible software package that has gained widespread use for untargeted metabolomic studies [15]. It is available through Bioconductor and can be installed in R using the following commands:

```
source("http://bioconductor.org/biocLite.R")
1.1.1.1. biocLite("xcms")
```

2.2. Data import and visualization

XCMS requires data in an open access nonproprietary format such as Network Common Data Form (NetCDF) or mzXML. File conversion often can be done within the operating software of the instrument. A popular online tool, ProteoWizard (http://proteowizard.sourceforge.net), is also available that can be used to convert raw LCMS data into an open data format. The converted data files are placed in a subdirectory named "cdf" within the R working directory. From here the data can be imported, processed, and visualized.

An LCMS data file is a series of successively recorded mass spectra over a range of *m/z* values. The total intensity of all ions at each time point is known as the total ion chromatogram (TIC). The *xcmsRaw* function is used to read data files into R's memory environment. The *plotTIC* function can be used to produce a TIC or base peak chromatogram (BPC). This function can also be used to obtain the numerical data for custom plotting. The following example demonstrates how to do this in R:

```
library(xcms) # load the xcms library
cdffiles <-list.files("./cdf", recursive=T, full=T) # define data
xr1 <- xcmsRaw(cdffiles[6]) # extract raw data
t1 <- xr1@scantime # extract time vector
int1 <- xr1@tic # extract TIC intensity vector
s1 <- rep("HG5", 1896) # create sample vector
g1 <- rep("hp-2dg Green", 1896) # create group vector
tic1 <- data.frame(t1, int1, s1, g1)
```

This script loads the xcms package, defines the raw data files, and creates an xcmsRaw object (xr1) from file number 6. This object stores information in the raw data that can be extracted and combined into a data frame. We add factor columns for sample and group that will be used to make a custom plot with ggplot2.

This process is repeated for additional raw data files. The individual data frames are combined into a single data frame with the *rbind* function.

```
tic.data <- rbind(tic1, tic2, tic3, tic4)
```

Once the data have been combined into a single data frame, a multipanel plot can be produced using the facet_wrap function in ggplot2. The result is shown in **Figure 2**. Variation in peak profiles can be readily observed between the different groups.

2.3. Raw data processing

XCMS uses several algorithms to process LCMS data. The first step is to filter and detect ion peaks using the *xcmsSet* method. The peak detection algorithm is based on cutting the data into slices of predefined mass widths and then finding peaks in the chromatographic time domain by applying a Gaussian model peak matched filter. Although the default arguments for the *xcmsSet* method may provide acceptable results in some cases, it is recommended that the parameters used for peak selection be optimized for this step. Without optimization, common problems that may arise include oversampling, i.e., assigning the same metabolite to multiple peaks, and missing values, i.e., failure to detect certain peaks in some samples that can interfere with downstream statistical analysis [8].

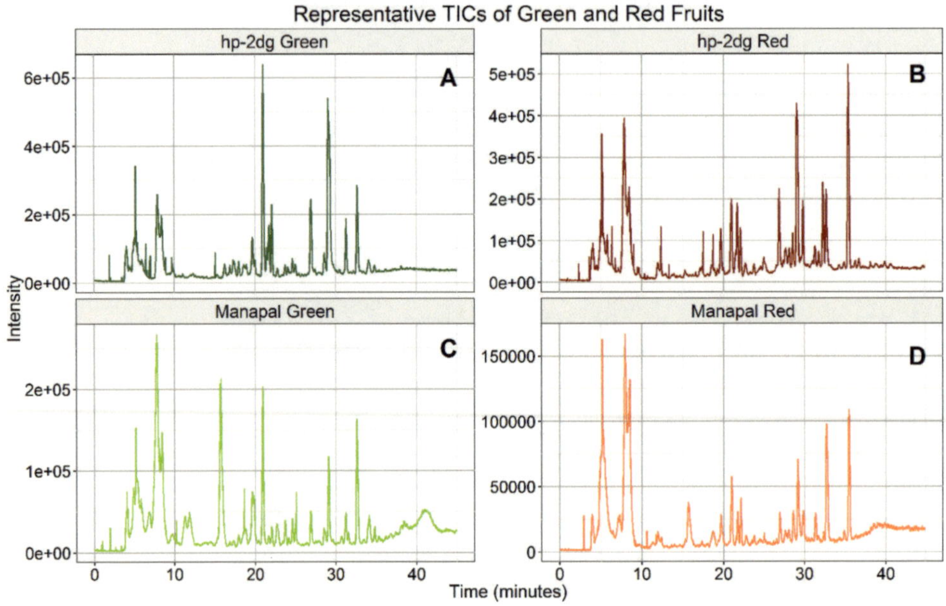

Figure 2. Representative total ion chromatograms (TICs) of green fruits (A, C) and red fruits (B, D) of the hp-2dg (A, B) and Manapal (C, D) tomato varieties.

The next step after filtration and peak identification is matching peaks across samples. Peaks representing the same analyte across samples must be placed into groups. This is accomplished with the *group* function, which returns a new xcmsSet object. After matching peaks into groups, XCMS can use those groups to identify and correct correlated drifts in retention time using the *retcor* function. The aligned peaks can then be used for a second pass of peak grouping which will be more accurate than the first.

After the second pass of peak grouping, there will still be missing peaks from some of the samples. This can occur because peaks were missed during peak identification or because an analyte was not present or below the detection limit in a sample. Missing values can be problematic for statistical methods that require a fully defined data matrix. Those missing data points can be filled in by re-reading the raw data files and integrating them in the regions of missing peaks using the *fillPeaks* function.

XCMS can generate a report showing fold change differences in analyte intensities and their statistical significance using the *diffreport* function. However, we recommend obtaining the raw peak integration results using the *groupval* function. This function returns a numerical matrix in which each row represents a peak defined by its mass and retention time and each column represents a different sample. In an example used here, 308 peaks were identified with 1.4% missing values. This matrix provides the starting point for downstream statistical analysis.

```
# Step 1.  Create xcmsSet object
xset <- xcmsSet(cdffiles, snthresh = 20, step = 0.5)
# Step 2.  Peak Alignment
xsg <- group(xset)
# Step 3.  Retention Time Correction
xsg2 <- retcor(xsg, method="obiwarp", plottype="deviation")
# Step 4.  Regroup and change bandwidth as desired
xsg3 <- group(xsg2, bw=10)
# Step 5.  Fill in missing peaks
xsg4 <- fillPeaks(xsg3)
# Step 6.  Get peak intensity matrix
results <- groupval(xsg4, "medret", "into")
# Step 7.  Save results as .csv file
write.csv(results, file="xcmsresults.csv")
```

3. Descriptive statistics

Once the raw data have been processed in XCMS, it is often useful to obtain descriptive statistics for each variable. This can be accomplished in R by creating a function to calculate the mean, median, standard deviation, standard error, and coefficient of variation for each variable. The *apply* function executes these operations on each row and returns the results into a data frame.

```
sumstats <- function(z) {
      Mean <- apply(z, 1, mean)
      Median <- apply(z, 1, median)
      SD <- apply(z, 1, sd)
      SE <- apply(z, 1, function(x) sd(x)/sqrt(length(x)))
      CV <- apply(z, 1, function(x) sd(x)/mean(x))
      result <- data.frame(Mean, Median, SD, SE, CV)
      return(result)
      }
```

This function creates a new data object containing descriptive statistics for each variable that can be used to rank variables according to mean or median intensity or to assess the degree of dispersion for each variable.

4. Data preprocessing

Before higher order statistical methods can be applied, it is often necessary to "clean up" the data to improve the analysis. Typical steps include (1) imputation of missing values, (2) transformation, (3) scaling, and (4) normalization.

4.1. Imputation of missing values

A common phenomenon in metabolomics measurements is the occurrence of missing values, i.e., empty cells where a respective metabolite peak has not been assigned any numerical value. As many multivariate methods require a fully defined matrix or become computationally inefficient for incomplete data, estimation of missing values is an important step in the preparation of the data [8].

Even after using the *fillPeaks* function in XCMS, there are still typically a large number of missing values. There are several strategies for dealing with these, including removing variables with missing values that exceed a certain threshold. However, these are often interesting features that are important in discriminating experimental groups. An alternative and widely used approach is imputation, where missing values are replaced with a small value with the assumption that the feature in question is below the limit of detection in those samples where XCMS fails to detect a peak.

The following function can be used to find the minimum nonzero value in a set of numbers and then divide that value by 2. We can then use the *apply* function to replace the missing values in each row of the matrix using this function.

```
replacezero <- function(x) "[<-"(x, !x | is.na(x), min(x[x > 0],
na.rm = TRUE) / 2)
newdata <- apply(data, 1, replacezero)
```

4.2. Log transformation

Since metabolomics studies are generally concerned with relative changes in metabolite levels, a log or other suitable transformation is normally applied before performing higher order statistical analysis. A log transformation helps to remove heteroscedasticity from the data and correct for a skewed data distribution [7]. This operation is easily performed in R using the *log* function. The default option is to compute the natural logarithm. However, the general form *log* (*x*, base) computes logarithms with any desired based. The base 2 log transformation is commonly used in metabolomics studies. Note that the *log* function will return NA for any zero values in the data matrix.

4.3. Normalization and scaling

The purpose of data normalization is to reduce systematic variation and to separate biological variation from nonbiological variation introduced by the experimental process. This is often necessary to improve the results of higher order statistical analysis [7, 23, 25].

Normalization can be sample wise or feature wise or both. Sample wise normalization makes the samples more comparable to each other. Common approaches include normalization to constant sum, to a reference sample or feature, or sample specific normalization such as dry weight or tissue volume.

http://dx.doi.org/10.5772/65405

Feature wise normalization involves centering the data around the mean combined with various types of scaling. Centering focuses the data on the amount of variation instead of the mean intensity. Scaling involves dividing each variable by a factor that approximates the amount of data dispersion. The most common scaling approach is known as unit scaling or autoscaling where each variable is mean centered and then divided by the standard deviation. After autoscaling all variables become equally important and are analyzed on the basis of correlations instead of covariances. A disadvantage of autoscaling is that it tends to inflate the importance of small variables which are more likely to contain measurement errors [7]. Other scaling operations include Pareto scaling, which uses the square root of the standard deviation as the scaling factor, vast scaling, which uses the standard deviation and the coefficient of variation as scaling factors, and range scaling, where the range is used as the scaling factor [7].

The *scale* function in R automatically performs centering and autoscaling. Other scaling procedures can also be carried out using the custom functions. For example, the following function can be used for Pareto scaling which is recommended for metabolomics data.

```
paretoscale <- function(z) {
       rowmean <- apply(z, 1, mean) # row means
       rowsd <- apply(z, 1, sd)  # row standard deviation
       rowsqrtsd <- sqrt(rowsd) # sqrt of sd
       rv <- sweep(z, 1, rowmean,"-")  # mean center
       rv <- sweep(rv, 1, rowsqrtsd, "/")  # divide by sqrtsd
       return(rv)
       }
```

5. Principal component analysis

Principal component analysis (PCA) is the foundation for many multivariate techniques that seek to describe a set of observations based on a large number of variables [25, 26]. The core idea of PCA is to reduce the dimensionality of the data, i.e., the number of variables while retaining as much of the variation as possible. Using PCA, it is possible to extract and display the systematic variation in the data and identify related groups, trends, and outliers.

PCA returns two important types of information: a scores matrix and a loadings matrix. The scores matrix contains the coordinates of the samples (i.e., observations) for each principal component and provides a summary of the observations in a lower dimensional space. The first principal component describes the largest variation in the data matrix, the second component describes the second largest, and so on. All PCA components are mutually orthogonal, meaning they are uncorrelated. Generally, most of the variation is captured in the first two or three principal components. Therefore, a scatter plot of the first two score vectors usually provides a good summary of all the samples and can reveal if there are differences between the groups as well as outliers.

Analogous to the scores matrix, the loadings matrix describes the relationships among the measured variables for each principal component. A scatterplot of the first two loading vectors can reveal the influence (weights) of individual variables in the model. An important aspect of PCA is that directions in the scores plot correspond to directions in the loadings plot, and so a comparison of these two plots can be used to identify, which variables (loadings) are most important for separating the different samples (scores) [28].

When there are more than two experimental classes, it is generally appropriate to use multivariate methods such as PCA to find patterns in the data [27]. The primary goal of these methods is to determine if the classes can be predicted from the variables (class assignment) and to identify which variables are important in predicting class membership.

There are several ways to perform PCA in R. Here, we will demonstrate the procedure using the *prcomp* function, which comes with the built-in R stats package. This method uses singular value decomposition (SVD) to calculate eigenvalues, which is the standard approach in PCA. The following syntax is used.

```
prcomp(data, retx = TRUE, center = TRUE, scale = FALSE)
```

where "data" is a dataframe or matrix containing the data, retx is a logical value that indicates whether the scores will be returned, center is a logical value indicating whether the variables should be mean centered, and scale is a logical value indicating whether the variables should be scaled to unit variance.

After missing values have been replaced, the data are log transformed and Pareto scaled. Note that the Pareto scale function must first be defined as above.

```
logdata <- log(newdata, 2)
pareto.logdata <- paretoscale(logdata)
```

Now, we perform PCA. The center and scale options are set to FALSE since these operations have already been performed with the *paretoscale* function.

```
pca <- prcomp(t(pareto.logdata), center=F, scale=F)
```

The *t* function is used here to transpose the matrix so that each row represents an observation (sample) and each column represents a variable (peak). This is necessary if we want the scores matrix to correspond to samples and the loadings matrix to correspond to variables.

The *prcomp* function returns several outputs that can be accessed by the *summary* command.

http://dx.doi.org/10.5772/65405

```
pcaresults <- summary(pca)
```

This returns a list that contains the standard deviations (eigenvalues) and proportion of the total variance for each principal component, the scores matrix, and the loadings matrix. We can extract each of these outputs into a new data frame and save the results to file for later use.

```
scree.data <- as.data.frame(pcaresults$importance)
score.data <- as.data.frame(pcaresults$x)
loadings.data <- as.data.frame(pcaresults$rotation)
write.csv(scree.data, "pca_scree.csv")
write.csv(score.data, "pca_scores.csv")
write.csv(loadings.data, "pca_loadings.csv")
```

5.1. Scores plot

The results of a PCA can be easily visualized using the base graphics functions in R. However, it is often desirable to produce a high-quality figure with custom formatting using ggplot2. To do this, we first import the scores matrix from the PCA. Since the first two principal components capture most of the variance, we will subset the data to include only those values.

```
data <- read.csv("pca_scores.csv", header=T)
data <- data[, c(1:3)] # subset columns 1-3
```

In order to map individual samples to their respective groups, we need to add a new column to the data frame indicating the group to which each sample belongs. To do this, we first create a vector of group names and then add the vector to the data frame with the *cbind* function. We can then generate the scores plot with ggplot2.

```
ggplot(data, aes(PC1, PC2)) +
geom_point(aes(shape=Group)) +
geom_text(aes(label=data$X)) +
stat_ellipse(aes(fill=Group))
```

This script defines the data and adds layers for data points, text labels, and confidence ellipses. The resulting plot is shown in **Figure 3**.

The scores plot shows that the two genotypes are well separated along the first PC axis while the developmental stage (green versus red) is separated along the second PC axis. There is more variation among the hp-2dg green samples than among the other groups.

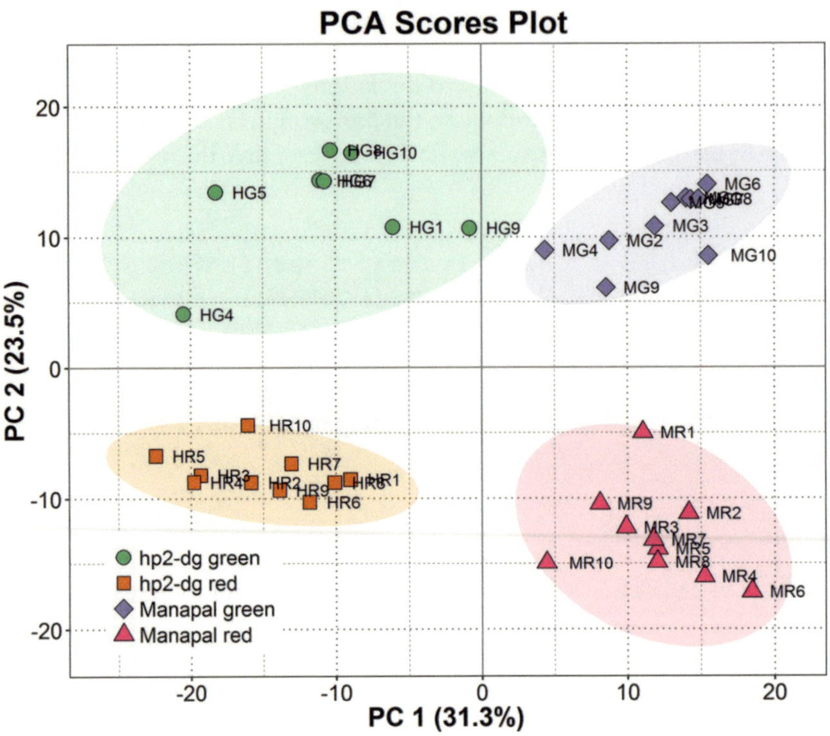

Figure 3. PCA scores plot.

5.2. Loadings plot

We can create a loadings plot using a similar approach. Since there are a large number of variables, we would also like to know which ones have the largest influence on the PCA. Variables with high loadings (positive and negative) are more likely to be important for discriminating groups that are separated in the scores plot.

One of the major advantages of R is that it has many powerful and flexible functions for subsetting data. One approach might be to identify the maximum and minimum loadings using the *range* function and then subset the data based on a percentage of these values. Alternatively, we can make a plot of PC1 versus PC2 loadings and visually inspect the data for high and low values. The *subset* function can be used to select rows that meet certain criteria. In this example, 0.09 and −0.09 were selected for threshold values.

```
significant.loadings <- subset(loadings, PC1 > 0.09 | PC1 < -0.09 |
                                         PC2 > 0.09 | PC2 < -0.09)
```

For plotting in ggplot2, it is generally recommended to add factor columns to the data frame for the purpose of mapping aesthetics to variables. A factor is a categorical value in R with predefined levels. We can use the *ifelse* function to specify the factor level in the new columns much in the same way as the *subset* function was used to create a new data frame. Loadings above and below the threshold values are marked for subsetting in this way.

```
loadings$pc1.change <-
  ifelse(loadings$PC1 > 0.09,"UP",
         ifelse(loadings$PC1 < -0.09,"DOWN",
                "nochange"))
```

With the added factors, we can make the plot in ggplot2 and indicate significant loadings with different colors. The grid package provides several options for adding text annotations. The resulting plot is shown in **Figure 4**.

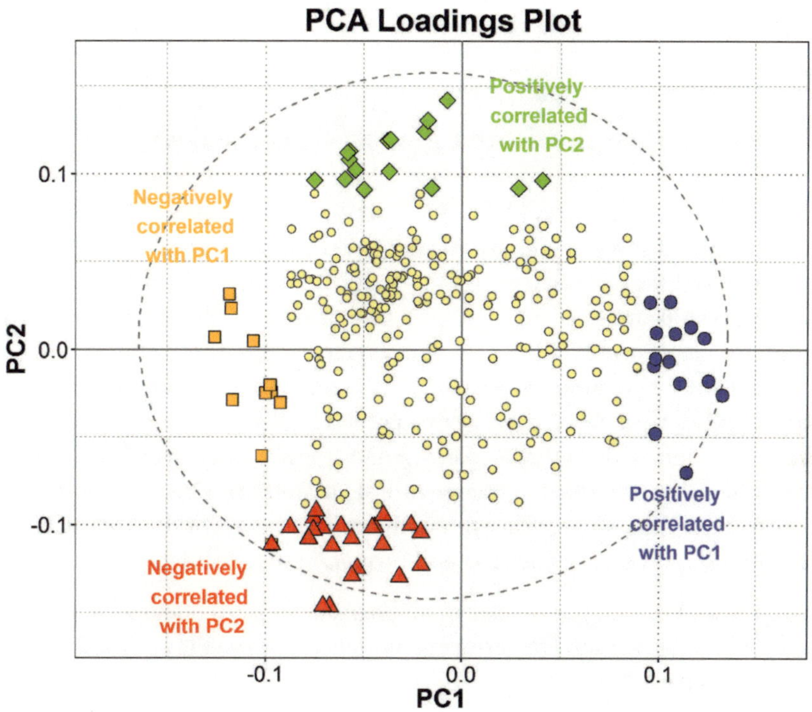

Figure 4. PCA loadings plot.

Since the scores matrix and the loadings matrix share similar geometric properties, the direction of the loadings indicate those variables that have the greatest influence on class separation. Based on these criteria, 64 potentially significant peaks were identified out of the original 308 (**Figure 4**). Separation along the PC1 axis identified features that show high variation by genotype while separation along the PC2 axis identified features that show high variation by developmental stage.

It should be emphasized that since PCA is an exploratory method, the interpretation of PCA results for the purpose of inferring biological importance must be done with caution. Potentially interesting features must be further analyzed to assess their biological significance. This can be done using boxplots, heatmaps, or other suitable graphical displays and rechecking the raw data by generating extracted ion chromatograms.

The first step in this process is to subset the original data to include only those variables of interest, i.e., the colored symbols in **Figure 4**. This can be done in R out using the extremely useful *merge* function.

```
data.sub <- merge(newdata, significant.loadings, by="row.names")
```

This command creates a new data frame containing the peak intensity values for the 64 variables with high PCA loadings that can be further analyzed as described below.

6. Partial least squares-discriminant analysis (PLS-DA)

Partial least squares-discriminant analysis (PLS-DA) is a supervised method that uses multiple linear regression to find the direction of maximum covariance between a data set and class labels [28]. Supervised techniques can be very helpful for highlighting sample/group differences when PCA results are masked by high levels of spectral noise, strong batch effects, or high within group variation [26]. PLS-DA sharpens the separation between groups of observations by rotating PCA components such that a maximum separation among classes is obtained and identifies variables that carry most of the class separating information. However, contrary to PCA, supervised methods like PLS-DA aggressively overfit models to the data, almost always yielding scores in which classes are separated [26, 27]. As a result, these methods can generate excellent class separation even with random data. For this reason results of these types of tests should be critically checked and properly cross-validated.

The pls, plsdepot, and muma packages can all be used for partial least squares analysis in R [33, 34]. We will demonstrate how to perform an extension of the PLS method known as OPLS-DA using the muma package below.

7. Heatmaps

Heatmaps are an effective tool for displaying feature variation among groups of samples [35]. The basic concept of a heatmap is to represent relationships among variables as a color image. Rows and columns typically are reordered so that variables and/or samples with similar profiles are closer to one another, making these profiles more visible. Each value in the data matrix is displayed as a color, making it possible to view the patterns graphically.

http://dx.doi.org/10.5772/65405

Heatmaps use an agglomerative hierarchical clustering algorithm to order and display the data as a dendrogram. Two important factors to consider when constructing a heatmap are the type of distance measure and the agglomeration method used. For details on the various methods available see [35].

The *heatmap.2* function in the gplots package can be used to construct a heatmap that is easily customizable and include options for both the distance and agglomeration methods, as well as scaling options for rows or columns. Unless the actual numerical values in the data matrix have an explicit meaning, row scaling is usually advisable [35].

A heatmap showing the scaled data from the 64 loadings extracted by PCA is shown in **Figure 5**. Four well-defined clusters are evident that correlate well with the four different experimental groups. These variables form a starting point for further experiments and analyses.

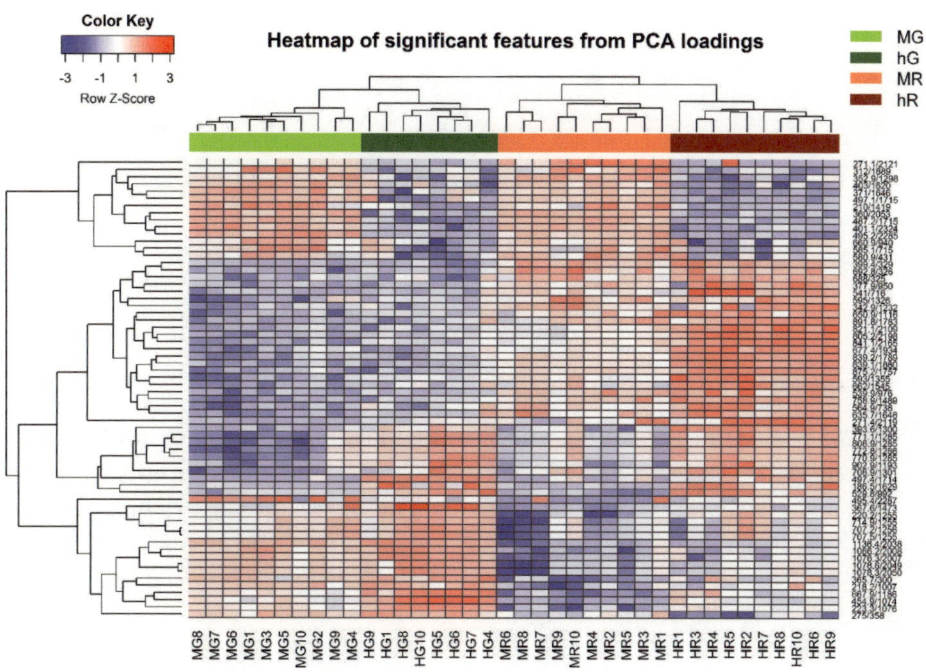

Figure 5. Heatmap of significant features obtained from PCA loadings.

8. Boxplots

Boxplots are another good way to visualize and compare features among different samples. A boxplot graphically depicts a vector through its five-number summary. The edges of the box represent the lower and upper quartiles while the whiskers represent the minimum and maximum values. The median is displayed as a line within the box. Outliers are represented as symbols outside of the whiskers.

A simple boxplot can be generated from any numeric vector using the *boxplot* function in R. However, a more customizable boxplot can be created using the ggplot2 package. **Figure 6** shows boxplots for four significant features from the PCA results. The data first were log transformed and Pareto scaled to show relative differences.

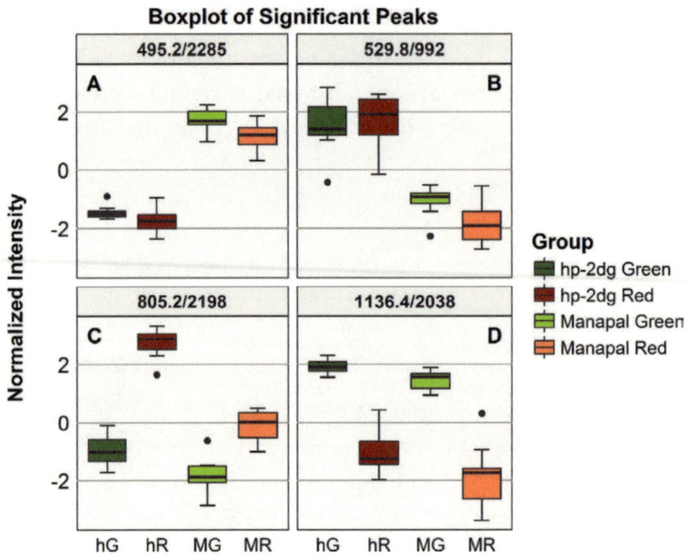

Figure 6. Boxplot of four significant peaks identified from PCA loadings.

The 495.2/2285 peak, which had a high positive PC1 loading, was significantly higher in the Manapal strain, whereas the 529.8/992 peak, which had a high negative PC1 loading, was significantly higher in the hp-2dg strain. The 1136.4/2038 peak, which had the highest positive loading for PC2, was significantly higher in green fruits of both varieties. Interestingly, the 805.2/2198 peak, which had a high negative loadings for both PC1 and PC2, was only significant in red fruits of the hp-2dg strain.

9. Extracted ion chromatograms

While statistical procedures provide important clues about potentially significant variables, a critical but often overlooked step in analyzing metabolomics data is reinspecting the raw data to assess the validity of these results. Not only can this provide confirmation of meaningful features but it can also reveal false positives caused by scaling artifacts or spurious peak assignment, which is common in XCMS-processed data.

LCMS ion peaks can be visualized through extracted ion chromatograms (EICs). An EIC is essentially a "slice" of the raw data that covers a specific *m*/*z* and time range. XCMS automat-

http://dx.doi.org/10.5772/65405

ically generates EICs for peaks that show high significance, but these are low quality "snapshot" images. However, the *plotEIC* function in XCMS can be used to extract the numerical data for any EIC of interest. The following commands describe how to obtain EIC data for all samples in a data set and generate an EIC plot for grouped samples.

We first create a list of xcmsRaw objects from the raw data files with the *lapply* function.

```
cdffiles <-list.files("./cdf", recursive = TRUE, full=T)
dat.raw <- lapply(cdffiles, xcmsRaw)
```

Next, we set the upper and lower limits for *m*/*z* and time. For this example, we will look at the 805.2/2198 peak since PCA and boxplots indicate that this peak was highly significant in red fruits of the hp-2dg strain.

```
mrange <- c(804.5, 805.5)
trange <- c(2000, 2400)
```

We can use the *lapply* function again to create an EIC for all samples. The data are then merged into a data frame for custom plotting in ggplot2.

```
eicraw <- lapply(dat.raw, plotEIC, mzrange=mrange, rtrange=trange)
eic.df <- do.call(rbind, lapply(eicraw, data.frame))
```

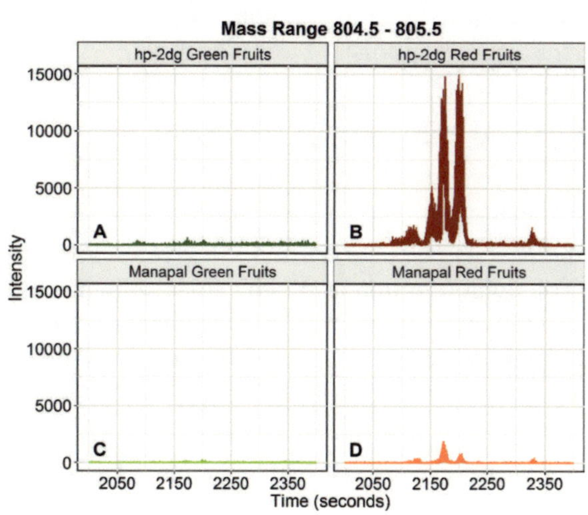

Figure 7. Extracted ion chromatograms for the *m*/*z* range 804.5-805.5 in hp-2dg green fruits (A), hp-2dg red fruits (B), Manapal green fruits (C) and Manapal red fruits (D). Each panel represents 10 samples.

The results are shown in **Figure 7**. The grouped EIC data clearly show that this feature is completely absent in green fruits and is very low in red fruits of Manapal. In contrast, there are several large peaks over this mass range in red fruits of hp-2dg, indicating this feature is a class-specific biomarker.

10. Volcano plots

A metabolomics experiment often involves a comparison of two groups, e.g., a treatment group versus a control. It is customary in such cases to use univariate methods to obtain a summary of the data and identify potentially important variables before applying multivariate methods [27]. A common tool to identify discriminatory features is to construct a volcano plot. This type of plot displays the fold change differences and the statistical significance for each variable. The log of the fold change is plotted on the x-axis so that changes in both directions (up and down) appear equidistant from the center. The y-axis displays the negative log of the p-value from a two-sample t-test. Data points that are far from the origin, i.e., near the top of the plot and to the far left or right, are considered important variables with potentially high biological relevance.

The steps required to construct a volcano plot can be carried out using several base R functions. The fold change is typically calculated as the ratio of the two means. We can use the *apply* function to determine the means for each variable.

```
group1mean <- apply(data[1:10], 1, FUN=mean)
group2mean <- apply(data[11:20], 1, FUN=mean)
```

We then divide the means of each variable to obtain the ratio and take the logarithm so that changes in both directions appear equidistant from the center.

```
FC <- group1mean/group2mean
log2FC <- log(FC,2)
```

The *t.test* function in the R stats package returns the p-value for an unpaired t-test of two independent samples. The default option is Welch's t-test, which assumes unequal variance. Note that data preprocessing steps, such as sum normalization and log transformation, are usually applied to make the samples more comparable and to reduce heteroscedasticity.

```
pvalue <- apply(log.data, 1, function(x)
         {t.test(x[1:10], x[10:20])$p.value } )
```

It is recommended to use a multiple testing correction when performing t-test on multiple variables. There *p.adjust* function in R provides several options for this, including the family

http://dx.doi.org/10.5772/65405

wise error rate (FWER), also known as the Bonferroni correction, and the false discovery rate (FDR), also known as the Benjamini-Hochberg correction. The false discovery rate is a less stringent condition than the family-wise error rate, so this method is preferred when one is interested in having more true positives.

```
pvalue.BHcorr <- p.adjust(pvalue, method = "BH")
```

We take the negative log10 values so that variables with low adjusted *p*-values (i.e., high significance) appear near the top of the plot.

```
pvalue.BHcorr.neglog <- -log10(pvalue.BHcorr)
```

Finally, the data are merged into a single data frame that can be plotted in ggplot2.

```
volcano.data <- data.frame(log2FC, pvalue.BHcorr.neglog)
```

A volcano plot comparison of the two tomato genotypes in green and red fruits is shown in **Figure 8**. Significant variables are shown as colored symbols. We selected rather conservative cutoff values of 2 and −2 for the log2 fold change (fold change >4) and 2 for the –log FDR adjusted *p*-value ($p < 0.01$) to highlight those features that showed the largest differences. In green fruits, this led to the identification of 38 metabolite peaks that were significantly higher in Manapal and 20 peaks that were higher in hp-2dg, while in red fruits, 40 metabolite peaks were higher in Manapal while 24 were higher in hp-2dg. As with the PCA loadings, these variables can be explored further with heatmaps, boxplots, EICs, etc.

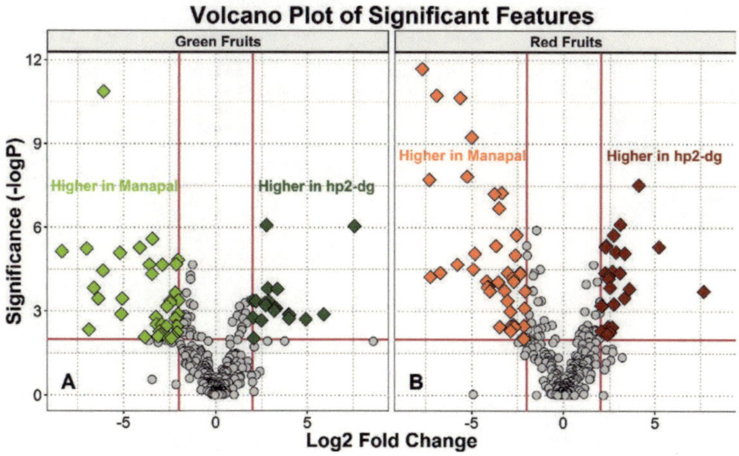

Figure 8. Volcano plot analysis of Manapal versus hp-2dg in green (A) and red (B) fruits.

The R package muma (Metabolomic Univariate and Multivariate Analysis) has a more sophisticated procedure for testing significance and returning *p*-values for a volcano plot [34]. Briefly, Shapiro Wilk's test for normality is performed to assess whether each variable has a normal distribution and to decide whether to perform a parametric test (Welch's *t*-test) or a nonparametric test (Wilcoxon-Mann Whitney test). The analysis returns fold change differences and a merged set of *p*-values from both tests and also applies a multiple testing correction that the user can specify. Finally, a volcano plot is generated highlighting significant variables based on the corrected *p*-values.

11. Orthogonal projection to latent structures-discriminant analysis (OPLS-DA)

An extension of the PLS technique known as orthogonal projection to latent structure is another very useful tool for analyzing metabolomics data. Like PLS this is a supervised method that pairs a data matrix ***X*** with a corresponding matrix ***Y*** containing sample information. The basic concept in OPLS is to separate the systematic variation in ***X*** into two parts, one that is correlated to ***Y*** and one that is not correlated (orthogonal) with ***Y*** [28]. Only the ***Y***-predictive variation is used to model the data. When working with discrete variables such as class labels the method is called OPLS discriminant analysis. The main advantages of OPLS-DA over PCA are better class discrimination and more robust identification of important features. The OPLS-DA algorithm normally is applied when there are only two classes comprising ***Y***.

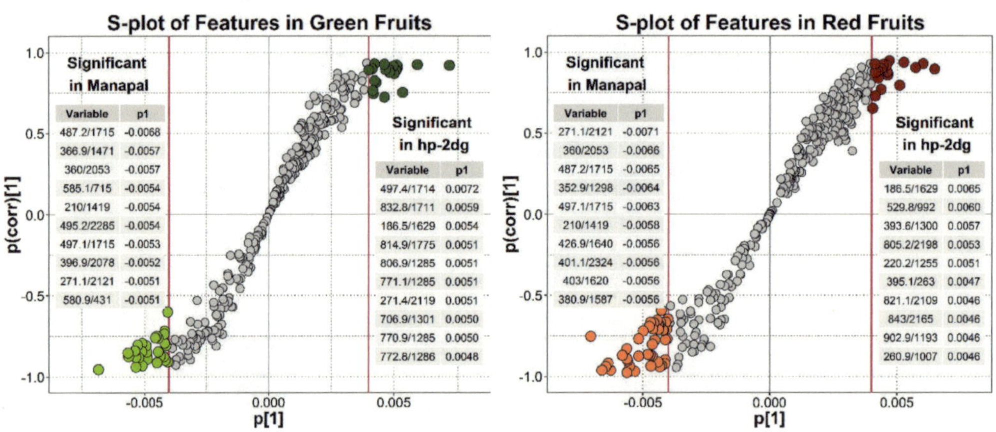

Figure 9. S-plots from OPLS-DA modeling of green and red fruits in Manapal and hp-2dg varieties.

The muma package can be used to perform OPLS-DA in R. A numerical class vector must be added to represent the ***Y*** matrix, i.e., the control group is given a value of 1 and the experimental group is given a value of 2. The data frame is saved in the working directory, and the analysis is carried out with two simple functions.

http://dx.doi.org/10.5772/65405

```
explore.data(file="logdata.csv", scaling="pareto")
oplsda(scaling="pareto")
```

This method automatically creates a new folder in the working directory that contains the OPLS-DA results in both numerical and graphical formats. The numerical data can be merged into a new data frame for custom plotting in ggplot2.

The loadings from an OPLS-DA model are displayed by means of an "S-plot" where the modeled covariance $p[1]$ is plotted on the x-axis and the correlation profile $p(\mathrm{corr})[1]$ is plotted on the y-axis. Variables with higher $p[1]$ values in both positive and negative directions have a larger impact on the variance between the groups, whereas variables with higher $p(\mathrm{corr})[1]$ values have more reliability. Therefore, data points that fall in the upper right and lower left quadrants have a high impact on the model and represent possible class-specific biomarkers.

Figure 9 shows S-plots from an OPLS-DA model of the two tomato varieties in green and red fruits. Variables with $|p[1]| > 0.004$ are highlighted, and the top 10 for each class are listed in tabular form on the graph. In general, there was good agreement between the OPLS-DA and volcano plot results for identifying significant variables.

12. Metabolite identification

One of the major challenges in LCMS-based metabolomics is metabolite annotation, i.e., identifying biological molecules from mass spectral data. Although metabolic profiling approaches that do not assign observed features to known metabolites can provide a powerful means of classifying and directly comparing samples, metabolite identification remains a crucial step for obtaining mechanistic insights into cellular processes. However, the complexity of the metabolome combined with the fact that many metabolites have not been structurally identified means that untargeted metabolomic studies typically yield a large number of unknown peaks.

The accurate identification of metabolites usually requires the ability to match candidate spectra with standard compounds run under the same conditions. Ideally, an orthogonal descriptor such as retention index is used for further validation. However, the lack of readily available standards remains a major obstacle in this regard, particularly in plant phytochemical studies [36]. Consequently, a number of strategies are being brought forward to assist in the chemical identification of unknown metabolites, including the development comprehensive mass spectral libraries [37–39], searchable databases [40–42], and information networks that integrate genomic, transcriptomic, and metabolomic data [43–45]. The construction, maintenance, and integration of these resources are crucial to the advancement of the field of metabolomics.

13. Conclusions

Untargeted metabolomics has become an increasingly powerful tool to investigate biological problems in agriculture, medicine, and a number of other fields. Therefore, efficient processing methods must be developed and refined to enable robust interpretation of metabolomics data. Method development and new software tools have helped address these challenges over the last decade. However, since improvements are still required at the various stages of data processing, establishing and refining new methods will continue to be important in the future of metabolomics research.

In this chapter, we have presented an overview of several common methods used for processing and analyzing LCMS-based metabolomics data and how to carry out these methods in the R programming environment. Although a variety of open source and web-based tools are available to support metabolomics data analysis, the ability to tailor the data processing workflow to one's own needs and generate custom graphics in R offers major advantages.

As the data sets used in all scientific disciplines get ever larger and more complex, it is becoming critical for scientists to be knowledgeable about how to use high-level languages such as R, which allow for easy and intuitive data manipulation. Along with powerful statistical capabilities, graphical tools make R an ideal environment for exploratory data analysis and provide exceptional flexibility for preparing high-quality publication-ready figures. Nevertheless, many technical and methodological issues must still be addressed to create analytical platforms that readily answer biological questions efficiently.

Acknowledgements

The authors like to thank Arthur Colvis for help with the LCMS experiments.

Author details

Stephen C. Grace* and Dane A. Hudson

*Address all correspondence to: scgrace@ualr.edu

Department of Biology, University of Arkansas, Little Rock, AR, USA

References

[1] Fiehn O. Metabolomics: the link between genotypes and phenotypes. Plant Mol Biol. 2002;48:155–171. DOI: 10.1023/A:1013713905833

[2] Kim HK, Verpoorte R. Sample preparation for plant metabolomics. Phytochem. Anal. 2010;12:4–13. DOI: 10.1002/pca.1188

[3] Pitt JJ. Principles and applications of liquid chromatography-mass spectrometry in clinical biochemistry. Clin. Biochem. Rev. 2009;30(1):19–34.

[4] Allwood JW, Goodacre R. An introduction to liquid chromatography-mass spectrometry instrumentation applied in plant metabolomic analyses. Phytochem. Anal. 2010;21:33–47. DOI: 10.1002/pca.1187

[5] Want EJ, Nordstrom A, Morita H, Siuzdak G. From exogenous to endogenous: the inevitable imprint of mass spectrometry in metabolomics. J. Proteome Res. 2007;6:459–468. DOI: 10.1021/pr060505+

[6] Schripsema J. Application of NMR in plant metabolomics: techniques, problems, and prospects. Phytochem. Anal. 2010;21:14–21. DOI: 10.1002/pca.1185

[7] van den Berg RA, Hoefsloot H, Westerhuis JA, Smilde AK, van der Werf MJ. Centering, scaling, and transformations: improving the biological information content of metabolomics data. BMC Genomics. 2006;7:142. DOI: 10.1186/1471-2164-7-142

[8] Steuer R, Morgenthal K, Weckwerth W, Selbig J. A gentle guide to the analysis of metabolomic data. In: Weckwerth W. Metabolomics: Methods and Protocols. Methods Mol. Biol. 2007;358:105–126. DOI: 10.1007/978-1-59745-244-1_7

[9] J.M. Quick. R Tutorial Series [Internet]. 2010. Available from: http://rtutorialseries.blogspot.com/

[10] Kabacoff RI. Quick-R: accessing the power of R [Internet]. 2014. Available from: http://www.statmethods.net/index.html

[11] Chang W. Cookbook for R [Internet]. Available from: http://www.cookbook-r.com/

[12] Sanderson A. R plot and graph gallery [Internet]. 2013. Available from: http://www.sr.bham.ac.uk/~ajrs/R/r-gallery.html

[13] Ross Z. Beautiful plotting in R: a ggpot2 cheatsheet [Internet]. 2016. Available from: http://zevross.com/blog/2014/08/04/beautiful-plotting-in-r-a-ggplot2-cheatsheet-3/

[14] Software and Programmer Efficiency Research Group. ggplot2 Quick Reference [Internet]. 2015. Available from: http://sape.inf.usi.ch/quick-reference/ggplot2

[15] Smith CA, Want EJ, O'Maille G, Abagyan R, Siuzdak G. XCMS: processing mass spectrometry data for metabolite profiling using nonlinear peak alignment, matching and identification. Anal. Chem. 2006;78:779–787. DOI: 10.1021/ac051437y

[16] Tautenhahn R, Patti GJ, Rinehart D, Siuzdak G. XCMS Online: a web-based platform to process untargeted metabolomic data. Anal. Chem. 2012;84:5035–5039. DOI: 10.1021/ac300698c

[17] Gowda H, Ivanisevic J, Johnson CH, Kurczy ME, Benton HP, Rinehart D, Siuzdak G. Interactive XCMS Online: simplifying advanced metabolomic data processing and

subsequent statistical analyses. Anal. Chem. 2014;86(14):6931–6939. DOI: 10.1021/ac500734c

[18] Xia J, Wishart DS. Web-based inference of biological patterns, functions and pathways from metabolomic data using MetaboAnalyst. Nat. Protocol. 2011;6:743–760. DOI: 10.1038/nprot.2011.319

[19] Xia J, Sinelnikov I, Han B, Wishart DS. MetaboAnalyst 3.0 - making metabolomics more meaningful. Nucleic Acids Res. 2015;43:251–257. DOI: 10.1093/nar/gkv380

[20] Grace SC, Embry S, Luo H. Haystack, a web-based tool for metabolomics research. BMC Bioinformatics. 2014;15(Suppl 11):S12. DOI: 10.1186/1471-2105-15-S11-S12

[21] Pluskal T, Castillo S, Villar-Briones A, Oresic M. MZmine 2: Modular framework for processing visualizing, and analyzing mass spectrometry-based molecular profile data. BMC Bioinformatics. 2010;11:395. DOI: 10.1186/1471-2105-11-395

[22] Lei Z, Li H, Chang J, Zhao PX, Sumner LW. MET-IDEA version 2.06: improved efficiency and additional functions for mass spectrometry-based metabolomics data processing. Metabolomics. 2012;8(S1):105–110. DOI: 10.1007/s11306-012-0397-5

[23] Katajamaa M, Orešič M. Data processing for mass spectrometry-based metabolomics. J. Chromatogr. A. 2007;1158:318–328. DOI: 10.1016/j.chroma.2007.04.021

[24] Sugimoto M, Kawakami M, Robert M, Soga T, Tomita M. Bioinformatics tools for mass spectroscopy-based metabolomic data processing and analysis. Curr. Bioinformatics. 2012;7:96–108. DOI: 10.2174/157489312799304431

[25] Scholz M, Selbig J. Visualization and analysis of molecular data. In: Weckwerth W. Metabolomics: methods and protocols. Meth. Mol. Biol. 2007;358:87–104. DOI: 10.1007/978-1-59745-244-1_6

[26] Worley B, Powers R. Multivariate analysis in metabolomics. Curr. Metabolomics. 2013;1:92–107. DOI: 10.2174/2213235X11301010092

[27] Saccenti E, Hoefsloot H, Smilde AK, Westerhuis JA, Hendriks MM. Reflections on univariate and multivariate analysis of metabolomics data. Metabolomics. 2014;10:361–374. DOI: 10.1007/s11306-013-0598-6

[28] Trygg J, Lundstedt T. Chemometrics techniques for metabonomics. The Handbook of Metabonomics and Metabolomics. J.C. Lindon, J.K. Nicholson, E. Holmes (eds.). Amsterdam, The Netherlands, Elsevier B.V. 2007; pp. 171–199.

[29] Levin I, Frankel P, Gilboa N, Tanny S, Lalazar A. The tomato dark green mutation is a novel allele of the tomato homolog of the DEETIOLATED1 gene. Theor. Appl. Genet. 2003;106:454–460. DOI: 10.1007/s00122-002-1080-4

[30] Peters JL, van Tuinen A, Adams P, Kendrick RE, Koornneef M. High pigment mutants of tomato exhibit high sensitivity for phytochrome action. Plant Physiol. 1989;134:661–666. DOI: 10.1016/S0176-1617(89)80024-0

[31] Mustilli AC, Fenzi F, Ciliento R, Alfano F, Bowler C. Phenotype of the tomato high pigment-2 mutant is caused by a mutation in the tomato homolog of DEETIOLATED1. Plant Cell. 1999;11:145–157.

[32] Bino RJ, Ric de Vos CH, Lieberman M, Hall RD, Bovy A, Jonker HH, Tikunov Y, Lommen A, Moco S, Levin I. The light-hyperresponsive high pigment-2dg mutation of tomato: alterations in the fruit metabolome. New Phytol. 2005;166:427–438. DOI: 10.1111/j.1469-8137.2005.01362.x

[33] Mevik BH, Wehrens R. The pls package: principal component and partial least squares regression in R. J. Stat. Softw. 2007;18:1–23. DOI: 10.18637/jss.v018.i02

[34] Gaude E, Chignola F, Spiliotopoulos D, Mari S, Spitaleri A, Ghitti M. muma, an R package for metabolomics univariate and multivariate statistical analysis. Curr. Metabolomics. 2015;1:180–189. DOI: 10.2174/2213235X11301020005

[35] Key M. A tutorial in displaying mass spectrometry-based proteomic data using heatmaps. BMC Bioinformatics. 2012;13(Suppl 16):S10. DOI: 10.1186/1471-2105-13-S16-S10

[36] Yang Z, Nakabayashi R, Okazaki Y, Mori T, Takamatsu S, Kitanaka S, Kikuchi J, Saito K. Toward better annotation in plant metabolomics: isolation and structure elucidation of 36 specialized metabolites from *Oryza sativa* (rice) by using MS/MS and NMR analyses. Metabolomics. 2014;10:543–555. DOI: 10.1007/s11306-013-0619-5

[37] Smith CA, O'Maille G, Want EJ, Qin C, Trauger SA, Brandon TR, Custodio DE, Abagyan R, Siuzdak G. METLIN: a metabolite mass spectral database. Ther. Drug Monit. 2005;27:747–751.

[38] Brown M, Dunn WB, Dobson P, Patel Y, Winder CL, Francis-McIntyre S, Begley P, Carroll K, Broadhurst D, Tseng A, Swainston N, Spasic I, Goodacre R, Kell DB. Mass spectrometry tools and metabolite-specific databases for molecular identification in metabolomics. Analyst. 2009;134:1322–1332. DOI: 10.1039/B901179J

[39] Kind T, Wohlgemuth G, Lee DY, Lu Y, Palazoglu M, Shahbaz S, Fiehn O. FiehnLib: mass spectral and retention index libraries for metabolomics based on quadrupole and time-of-flight gas chromatography/mass spectrometry. Anal. Chem. 2009;81:10038–10048. DOI: 10.1021/ac9019522

[40] Wishart DS, Knox C, Guo AC, Eisner R, Young N, Gautam B, Hau DD, Psychogios N, Dong E, Bouatra S, Mandal R, Sinelnikov I, Xia J, Jia L, Cruz JA, Lim E, Sobsey CA, Shrivastava S, Huang P, Liu P, Fang L, Peng J, Fradette R, Cheng D, Tzur D, Clements M, Lewis A, De Souza A, Zuniga A, Dawe M, Xiong Y, Clive D, Greiner R, Nazyrova A, Shaykhutdinov R, Li L, Vogel HJ, Forsythe I. HMDB: a knowledgebase for the human metabolome. Nucleic Acids Res. 2009;37:D603–D610. DOI: 10.1093/nar/gkn810

[41] Sawada Y, Nakabayashi R, Yamada Y, Suzuki M, Sato M, Sakata A, Akiyama K, Sakurai T, Matsuda F, Aoki T, Hirai MY, Saito K. RIKEN tandem mass spectral database

(ReSpect) for phytochemicals: A plant-specific MS/MS-based data resource and database. Phytochemistry. 2012;82:38–45. DOI: 10.1016/j.phytochem.2012.07.007

[42] Udayakumar M, Prem Chandar D, Arun N, Mathangi J, Hemavathi K, Seenivasagam R. (2012). PMDB: Plant metabolome database-a metabolomic approach. Med. Chem. Res. 2012;21(1):47–52. DOI: 10.1007/s00044-010-9506-z

[43] Bais P, Moon SM, He K, Leitao R, Dreher K, Walk T, Sucaet Y, Barkan L, Wohlgemuth G, Roth MR, Wurtele ES, Dixon P, Fiehn O, Lange BM, Shulaev V, Sumner LW, Welti R, Nikolau BJ, Rhee SY, Dickerson JA. PlantMetabolomics.org: a web portal for plant metabolomics experiments. Plant Physiol. 2010;152:1807–1816. DOI: 10.1104/pp.109.151027

[44] Sucaet Y, Wang Y, Li J, Wurtele ES. MetNet Online: a novel integrated resource for plant systems biology. BMC Bioinformatics. 2012;13:267. DOI: 10.1186/1471-2105-13-267

[45] King ZA, Lu J, Dräger A, Miller P, Federowicz S, Lerman JA, Ebrahim A, Palsson BO, Lewis NE. BiGG Models: A platform for integrating, standardizing and sharing genome-scale models. Nucleic Acids Res. 2016;44(D1):D515–D522. DOI: 10.1093/nar/gkv1049

Section 2

Metabolomics in Plant Metabolism and Agriculture

Chapter 5

^{13}C-Isotope-Labeling Experiments to Study Metabolism in *Catharanthus roseus*

Qifang Pan, Natali Rianika Mustafa,
Robert Verpoorte and Kexuan Tang

Additional information is available at the end of the chapter

http://dx.doi.org/10.5772/65401

Abstract

Plant metabolism is a complex network. Pathways are correlated and affect each other. Secondary metabolic pathways in plant cells are regulated strictly, and upon an intra- or extra-stimuli (e.g. stress), the metabolic fluxes will change as a response on the stimuli, for example, to protect the plant against herbivores or against microbial infections. ^{13}C-isotope-labeling experiment has been performed on cell cultures and hairy roots of *Catharanthus roseus* to measure fluxes through some pathways. However, due to the complexity of the total metabolic network in an intact plant, no experiments have yet been carried on *C. roseus* plants. In this study, [1-^{13}C] glucose was first applied to *C. roseus* seedlings grown in Murashige and Skoog's (MS) medium. In a time course, the amount and position of ^{13}C incorporation into the metabolites were analyzed by proton nuclear magnetic resonance (^{1}H NMR) and ^{1}H-^{13}C heteronuclear single quantum coherence (HSQC) NMR. The results show that the fed ^{13}C-isotope was efficiently incorporated into and recycled in metabolism of the intact *C. roseus* plant. The *C. roseus* plants seem to be a good system for metabolic flux analysis.

Keywords: ^{13}C-isotope labeling, *Catharanthus roseus*, intact plant, metabolic fluxes, NMR, HSQC

1. Introduction

Metabolic flux analysis (MFA) aims at the quantitation of the carbon flow through a metabolic network by measuring the enrichment and position of labels in the various measurable metabolites after feeding a labeled precursor *in vivo* or *in vitro*. Though now common in microorganisms, in plants, with their complex secondary metabolic pathways; MFA is so far mostly focused on primary metabolism. In fact, each metabolic flux reflects the function and performance of a specific pathway in a plant's development and its interaction with the environment, for example, defense against herbivores or microorganisms [1]. Consequently, metabolic fluxes represent the fourth dimension of a living organism. There are three dimensions of space, which form the phenotype, but the dynamics, the fluxes, represent life. Flux analysis based on ^{13}C-isotope-labeling experiments (^{13}CLE) has been established as an effective method for determining the flux distribution through the compartmented pathways of primary metabolism in plant cells. The ^{13}C isotope is not radioactive, thus convenient to be used to label the metabolites in the pathways. Usually, a specifically ^{13}C-isotope-labeled substrate, for example, [1-^{13}C] glucose, is used in a CLE study. After feeding, this labeled material is distributed over the various metabolic pathways. At various time points, the distribution of the label over the various measurable metabolites is measured by using different NMR or MS instruments [2, 3]. By NMR, the position of the label as well as the enrichment on every position in a molecule can be measured. By MS, the overall enrichment of a molecule can be determined, whereas the position can only to some extent be determined by analysis of the fragments.

There are two strategies for ^{13}C MFA: one is dynamic-labeling strategy (time course experiments), and the other is steady-state-labeling strategy. The dynamic-labeling strategy has an advantage in studying small partial networks and it is particularly effective for the analysis of secondary metabolism [1]. In this approach, a specific labeled advanced precursor of a pathway is pulse fed, and after a given time the incorporation is measured in the products of the pathway involved. In a steady-state-labeling strategy, the organisms are permanently growing in a medium containing a very early substrate for primary metabolism (e.g., a labeled sugar of pyruvate) and the diffusion of the label through all pathways is monitored by measuring the incorporation and position of the label in all measurable metabolites. This approach is usually utilized in studies of central carbon metabolism. In fact, the limiting factor in flux analyses in plants is the detection limits for the various metabolites, as the level of primary metabolites in plants is many folds higher than of secondary metabolites, the dynamic range analytical tools hamper the analysis of minor compounds. Therefore, often specific and selective extraction methods are used for the dynamic approach, whereas for the steady-state approach one uses the more general metabolomics analytical protocols.

In *Catharanthus roseus*, ^{13}C label has been applied for both pathway elucidation and system-wide flux quantification. By feeding [1-^{13}C] glucose to a cell culture of *C. roseus* with ^{13}C NMR spectroscopy [4], the biosynthetic pathway of secologanin was elucidated from which secologanin was found to originate from the triose-phosphate pathway. Salicylic acid biosynthesis was uncovered in *C. roseus* cell cultures by a retrobiosynthetic study based on ^{13}C feeding experiments [5]. Flux quantification in central carbon metabolism of *C. roseus* hairy roots by ^{13}C-labeling-based flux analysis, and quantitative assessment of crosstalk between the two isoprenoid biosynthesis pathways in cell cultures of *C. roseus* were also reported [6, 7]. Antonio et al. (2013) used plant cell suspension cultures of *C. roseus* to study the changes in fluxes after elicitation with jasmonate. The incorporation of fully labeled pyruvate was measured by gas chromatography-mass spectrometry (GC-MS) and ultra performance liquid chromatography (UPLC)-MS. The elicitation was found to disturb various metabolic pathways, as could be concluded from the differences in the incorporation of labels. Up to now, ^{13}CLE-based MFA has not been implemented on intact *C. roseus* plants. The major reason is that intact plants are a more complex metabolic

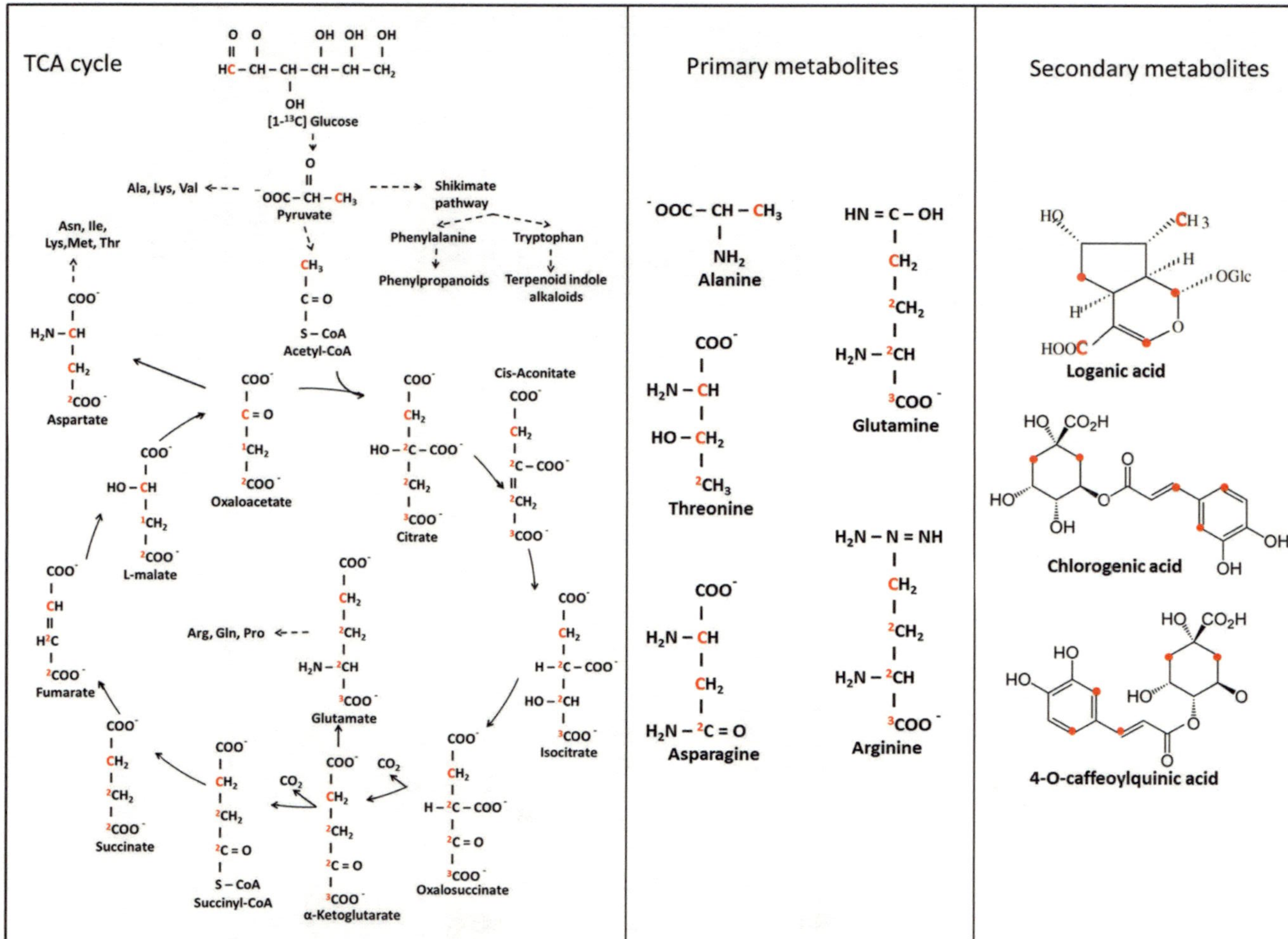

Figure 1. Label pattern of TCA cycle, amino acids, loganic acids, chlorogenic acid, and 4-*O*-caffeoyl quinic acid based on [1-^{13}C] glucose feeding. Carbons and filled cycles in red indicate ^{13}C-enriched positions.

system than cell cultures or hairy root cultures which only have a few different cell types. For example, previous research showed that some valuable TIAs (e.g., vindoline, vinblastine, and vincristine) can only be produced in leaves of *C. roseus*, not in cell cultures and hairy roots, due to the tissue- and cell-specific organization of terpenoid indole alkanoid (TIA) biosynthesis. So, a more detailed understanding of carbon flux distribution in the complex metabolic networks of intact *C. roseus* plants is a prerequisite for progress in metabolic engineering of TIA production in order to feed the rapidly growing market demands of these important TIAs.

In this study, the fate of [1-^{13}C] glucose fed to the intact *C. roseus* plants via the root system was analyzed in considerable detail. Labeling patterns of targeted metabolites were deduced from previous publications [4, 5, 8] (**Figure 1**), and confirmed by the current experiment. By tracing the label in some of the detected primary and secondary metabolites through a time course, we have information about the ^{13}C incorporation status of these compounds and thus in the metabolic fluxes in the *C. roseus* plant metabolism and the channeling of carbon into the MIA biosynthesis. Also, the metabolic changes after elicitation were measured in this model.

2. Materials and methods

2.1. Plant material and in vitro culture

C. roseus seeds (Pacifica Cherry Red cultivar) were purchased from PanAmerican Seed Company (USA). The seeds were surface sterilized in 75% of ethanol (v/v) and 10% of NaClO (v/v) for 1 and 15 min (respectively), and subsequently washed three times with sterile distilled water. Sterilized seeds were germinated on solid MS [9] basal medium with 1% non-labeled glucose. Finally, 54 seedlings were obtained, subcultured every 2 weeks in the same MS solid medium, and used before flowering for the labeling experiments. After 10 weeks, 19 plants as control were transferred to glass tubes and reared (each) with 5 ml of 10 g/l non-labeled glucose solution, whereas the other 35 plants were placed in separate glass tubes containing 5 ml of 10 g/l [1-^{13}C] glucose solution. The plant cultures were grown in a climate chamber under a 16-h light and 8-h dark photoperiod at 25 ± 2°C.

2.2. Chemicals

Murashige & Skoog medium (including vitamins) and gelrite (strength: 550–850 g/cm^2) were purchased from Duchefa Biochemie. D (+)-Glucose (>99.0%) was obtained from Fluka Chemie (Buchs, Germany), whereas [1-^{13}C]-D-glucose (>99.0%, with > 99% atom 1-^{13}C) was from Campro Scientific (Veenendaal, The Netherlands). Jasmonic acid (JA) was from Sigma-Aldrich Chemie (Steinheim, Germany).

2.3. Jasmonic acid elicitation

A stock solution of 10 mg/ml of JA 40% EtOH was prepared, filter sterilized, and used for elicitation. After 5 days of submerging the plant roots with 5 ml of [1-^{13}C] glucose solution (1%, w/v), 11 μl of the JA stock solution was aseptically spiked into each tube. The control samples received only the same volume of 40% EtOH. The plants were harvested at 0, 6, 24, and 72 h after treatment; young leaves, old leaves, stems, and roots of *C. roseus* plants were harvested separately, immediately frozen and ground in liquid nitrogen into powder and freeze-dried for 72 h before NMR extraction (five replicates per sample).

2.4. NMR analysis

^{1}H NMR spectra were recorded in CH_3OH-d_4 using a Bruker DMX 600-MHz spectrometer, while the coherence-order selective gradient heteronuclear single quantum coherence (HSQC) spectra were recorded in CH_3OH-d_4 using a Bruker AV 500-MHz spectrometer. HSQC spectra were recorded for a data matrix of 256 × 2048 points covering 30182.7 × 7812.5 Hz with 64 scans for each increment [10]. INEPT transfer delays were optimized for a heteronuclear coupling of 145 Hz and a relaxation delay of 1.5 s was applied. Data were linear predicted in F_1 to 512 × 2048 using 32 coefficients and then zero-filled to 2048 × 2048 points prior to echo-anti-echo-type two-dimensional (2D) Fourier transformation and a sine bell-shaped window function shifted by $\pi/2$ in both dimensions was applied. One-dimensional (1D) projection along the F_1-axis was extracted using the build-in positive projection tool of Topspin (version 2.1, Bruker Biospin).

The signal intensity of carbons at certain positions of a given metabolite was obtained from peak height in the ^{13}C-dimension spectra abstracted from the 2D HSQC spectra. The signal height of CH_3OH-d_4 was selected as the standard and set as 1 in both labeled and non-labeled samples. The other signals were normalized and expressed relative to this signal. ^{13}C signal intensity ratio was calculated by comparison of normalized ^{13}C signal heights between ^{13}C-labeled and non-labeled samples.

3. Results and discussion

3.1. Comparison of growth and metabolism of *C. roseus* plants grown in the solid MS medium versus soil

Two batches of *C. roseus* seeds (each containing of 10 seeds) were germinated, one batch in soil and another one in solid MS medium with glucose. They were kept in the same condition of light and temperature. The height, the size, and the leaf pairs of the plants from seedling until flowering were monitored and recorded regularly to determine the growth state of plants.

After 10–12 days, seeds in both batches germinated and produced their first pair of leaves. In the first 3 weeks after germination, there were no significant differences of height, leaf pairs, and leaf size between plantlets grown in MS medium and in the soil (**Figure 2**). However, in the following days, the plantlets in MS medium provided one more pair of leaves than those in soil did, but the leaf size was much smaller than that of plantlets grown in the soil (**Figure 2A** and **B**). Moreover, the soil plantlets grew higher than those grown in MS medium (**Figure 2C**). Plantlets in MS medium entered flowering time around 100 days after sowing, whereas those in soil flowered at 75 days. The plantlets grown in soil had a higher growth rate and a larger biomass than those grown in MS medium.

Metabolic differences between the plants grown in soil and MS medium were observed by ^{1}H NMR (**Figure 3**). The ^{1}H NMR spectra showed that qualitatively metabolites of plants grown

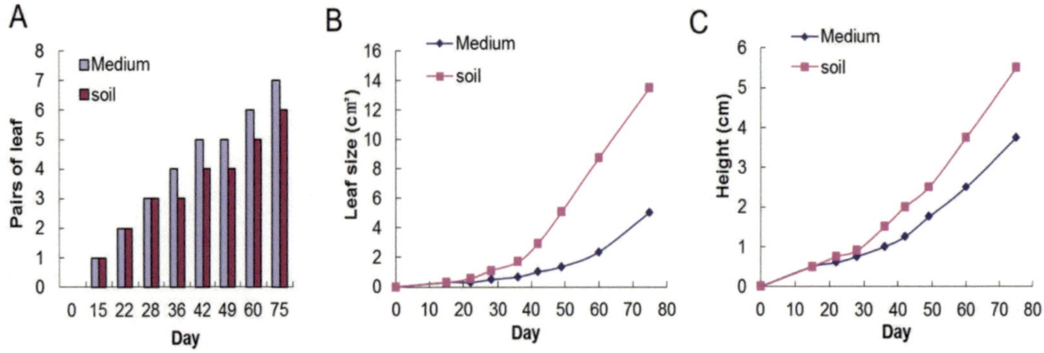

Figure 2. Comparison of the number of leaf pairs (A), leaf size (B), and height (C) of *Catharanthus roseus* plants grown in MS medium and soil during the development stage.

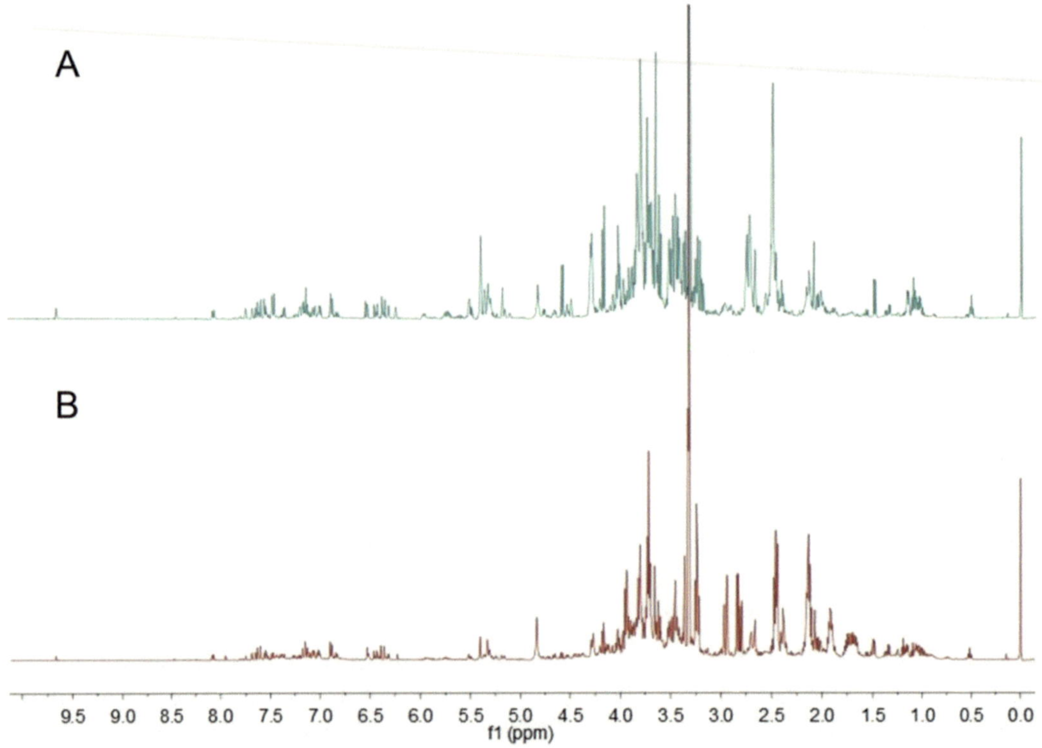

Figure 3. ^{1}H NMR spectrum of the crude extracts of *Catharanthus roseus* plants grown in soil (A) and MS medium (B).

in soil or MS medium were similar, but the levels varied (**Table 1**). Plants grown in soil produced higher levels of organic acids and sugars (malate, fumaric acid, glucose, and sucrose) than those grown in MS medium, indicating a low function/reduced level of carbon fixation in the leaves of the MS-grown plants. Also, secondary metabolites (such as secologanin, vindoline, quercetin, and kaempherol) were found in higher levels in soil-grown plants than the

http://dx.doi.org/10.5772/65401

Compounds	Signal intensity		Signal intensity ratio (S/M)
	Soil (S)	Medium (M)	
Vindoline	0.42	0.22	1.91
Threonine	0.27	0.34	0.79
Lactic acid	0.14	0.22	0.64
Alanine	0.54	0.51	1.06
Arginine	0.82	4.27	0.19
Quinic acid	0.21	0.30	0.70
Glutamate	1.25	2.14	0.58
Glutamine	2.94	7.08	0.42
Malate	5.96	1.82	3.27
Asparagine	0.03	0.45	0.07
β-glucose	0.67	0.13	5.15
α-glucose	0.42	0.08	5.25
Sucrose	1.34	0.39	3.44
Chlorogenic acid	0.12	0.11	1.09
Fumaric acid	0.10	0.07	1.43
Catharanthine	0.20	0.17	1.18
4-*O*-Caffeoyl quinic acid	0.15	0.12	1.25
Quercetin-3-*O*-glucoside	0.10	0.04	2.50
Kaempherol	0.13	0.08	1.63
Secologanin	0.13	0.03	4.33

Table 1. Comparison of metabolite levels in *Catharanthus roseus* plants grown in soil and MS medium, based on ^{1}H NMR.

plants grown in MS medium. On the other hand, plants cultured in MS medium displayed significantly higher levels of arginine, glutamine, and asparagine but relatively low level of glucose and sucrose. The levels of threonine, glutamate, quinic acid, and lactic acid were also higher in plants grown in MS medium than those in soil.

Some groups of metabolites have a close correlation with plant growth and biomass, such as the tricarboxylic acid cycle (TCA) intermediates succinate, citrate, or malate, as well as amino acids [11]. Both glutamine and asparagine are the major compounds for nitrogen fixing, transport, and storage in plants [12]. With the much more abundant nitrogen source in the medium than in the soil, the high levels of the amino acids in the medium grown plants could be explained. Meyer et al. [11] reported a negative correlation to the plant biomass with glutamine, which is in line with our findings. Sucrose starvation may lead to the presence of a large excess of asparagine in plant cells [13]. In the present study, the plants cultured on solid

MS medium require an aseptic jar with a cap, which limits the space to grow, and also affects air exchange, CO_2 availability, and accumulation of volatiles in the head space if compared with plants grown in soil. Despite the uptake of carbohydrates from the medium through the roots, the growth was less than the plants grown in soil which are dependent of carbon fixation by leaves. The limited availability of CO_2 in the sterile closed containers may thus be a reason for lower biomass production.

3.2. [1-^{13}C] glucose feeding experiment and JA elicitation on *C. roseus* plantlets

Samples from different organs (upper and lower leaves, stems, and roots) were measured by proton and carbon NMR. After feeding the plants with [1-^{13}C] glucose for 5 days, the incorporation of ^{13}C label was found in some primary and secondary metabolites detected in all organs of the *C. roseus* plantlets. ^{13}C signals of some primary and secondary metabolites were assigned based on the "in-house" database and some references [14, 15], and confirmed in Chapter 4. Totally, 14 amino acids, nine organic acids, two carbohydrates, six phenylpropanoids, five TIAs, two terpenoids, and three other compounds were identified. Among them, only the metabolites from which characteristic signals were clearly visible and non-overlapping in both ^{1}H and ^{13}C NMR spectra were quantified (**Figure 4**). Those include some primary metabolites such as amino acids (threonine, alanine, asparagine, aspartate, glutamine, glutamate, and arginine) and malic acid (**Figure 4A**), as well as some secondary metabolites such as phenylpropanoids (chlorogenic acid and 4-*O*-caffeoyl quinic acid), terpenoids (loganic acid and secologanin), and TIA (vindoline) (**Figure 4C**).

Figure 5 shows the ^{13}C-dimension HSQC spectra and ^{1}H NMR spectra of the non-labeled sample and the ^{13}C-enriched sample determined in CH_3OH-d_4. As expected, the superimposed ^{1}H-NMR spectra of leaves (**Figure 5**) and stems (data not shown) did not show any significant difference in proton signal intensity of the metabolites for the control and the ^{13}C-enriched sample. Production of these sugars caused a decrease in the levels of glucose and sucrose in roots but did not affect the metabolite levels in other organs. Except this, there was no significant change in metabolite levels of the plants fed with labeled- and non-labeled glucose solution (**Table 2**). This information is necessary to confirm that the ^{13}C signals of the spectra of enriched samples are due to incorporation of label, and not because of higher levels of production of the metabolites. Superimposed ^{13}C-dimension HSQC spectrum showed that the enriched sample had a much higher intensity of ^{13}C signals than the non-labeled one. The results indicate that the [1-^{13}C] glucose fed *C. roseus* plants grew normally, and incorporated the labeled sugar into its metabolic network. Previous work with *Arabidopsis* supports that ^{13}C feeding does not in itself distort the fluxes through the metabolic network in a plant [16].

3.3. ^{13}C incorporation into primary and secondary metabolites

The signals in the HSQC spectra of the enriched samples were identified (**Figure 4**). The carbon position of ^{13}C incorporation into a metabolite was investigated by calculating ^{13}C signal intensity ratios between the same carbons of the metabolite in labeled and non-labeled samples (**Table 3**).

http://dx.doi.org/10.5772/65401

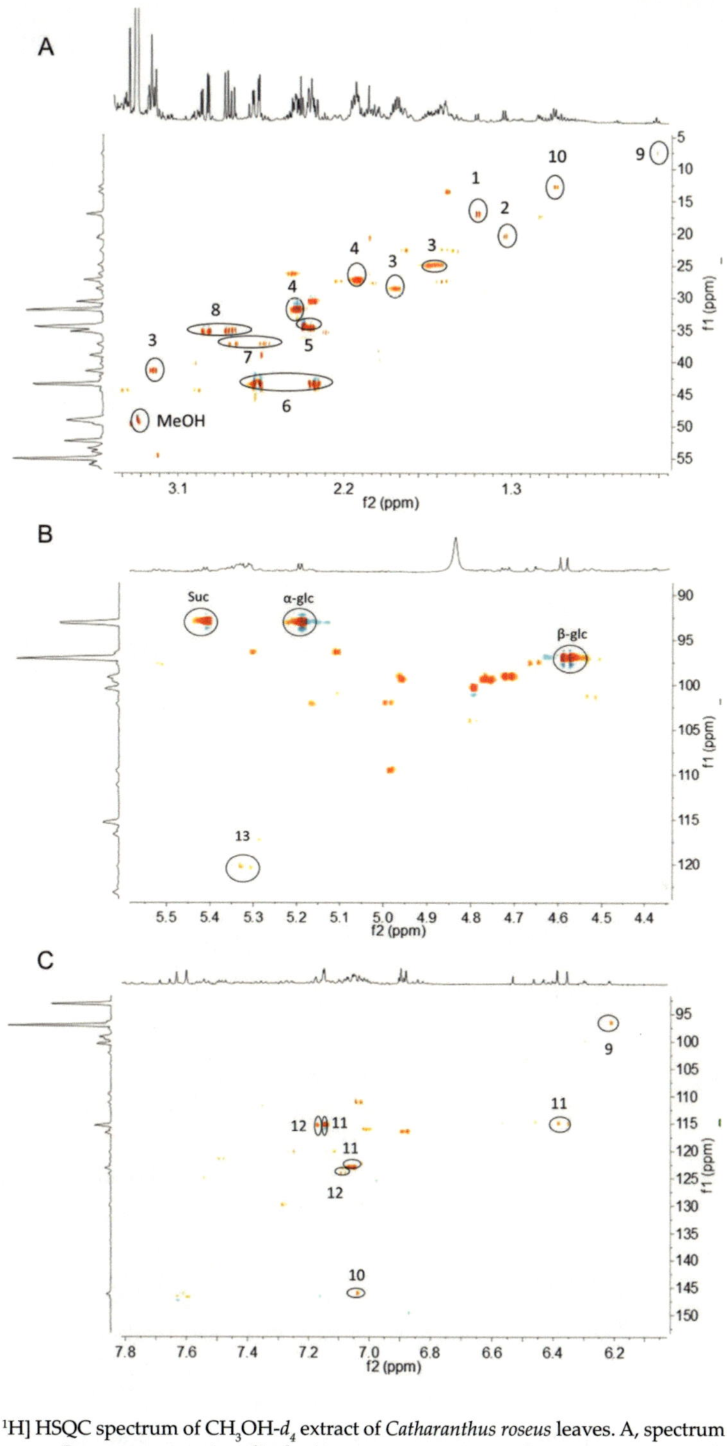

Figure 4. 2-D [^{13}C, ^{1}H] HSQC spectrum of CH_3OH-d_4 extract of *Catharanthus roseus* leaves. A, spectrum region displaying amino acid resonances; B, spectrum region displaying sugar resonances; C, spectrum region displaying aromatic resonances. 1, alanine; 2, threonine; 3, arginine; 4, glutamine; 5, glutamate; 6, malate; 7, aspartate; 8, asparagine; 9, vindoline; 10, loganic acid; 11, chlorogenic acid; 12, 4-*O*-caffeoyl quinic acid.

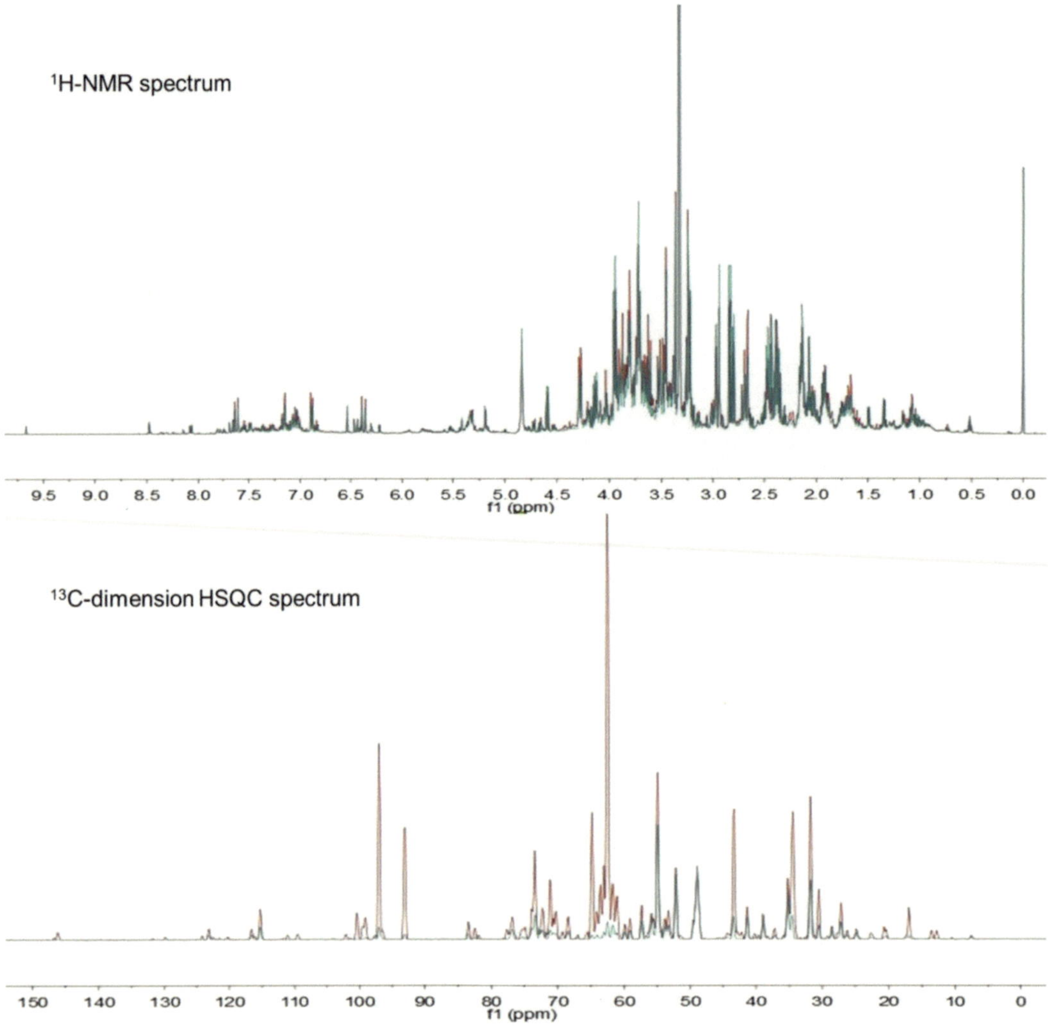

Figure 5. Superimposed ^{1}H NMR spectra and ^{13}C-dimension HSQC spectrum of labeled and non-labeled *Catharanthus roseus* plants. Spectra in green were non-labeled plants sample, spectra in red were ^{13}C-labeled plants sample.

Among amino acids, the signals corresponding to C at δ 16.98, C-3 of alanine, exhibited a high ^{13}C relative enrichment ratio. Glycolysis introduces the C-1 or C-6 of glucose into alanine C-3 [8]. Carbon signals at δ 20.47 of threonine and at δ 37.21 of aspartate also showed a relatively high labeling. The carbons of arginine and asparagine were apparently less labeled.

Glutamate (C-3 at δ 27.74, C-4 at δ 34.44, and C-5 at δ 55.67) and glutamine (C-3 at δ 27.11, C-4 at δ 31.83, and C-5 at δ 55.02) showed clear high ^{13}C incorporation. The relative enrichment ratios of C-3 and C-2 of glutamine were lower than that of C-4, which indicate the entry of a diluting flux of C4 compounds into the TCA cycle [17]. For glutamate, however, C-4 had a lower relative enrichment ratio than C-3 and C-2. Non-symmetrical enrichment ratios of

Compounds	Ratio of metabolite levels in labeled and non-labeled samples, (L0/C0), based on ^{1}H NMR			
	Upper leaf	Lower leaf	Stem	Root
β-Glucose	1.04	0.76	1.01	0.42
α-Glucose	0.97	0.67	1.03	0.39
Sucrose	0.95	0.62	0.82	0.65
Threonine	1.25	0.97	0.88	1.05
Alanine	1.12	0.92	0.96	0.82
Arginine	1.16	0.75	0.82	0.74
Glutamate	1.13	0.95	0.89	0.70
Glutamine	0.78	0.60	0.89	0.81
Aspartate	1.28	0.83	0.87	1.22
Asparagine	0.88	0.85	0.77	0.93
Malic acid	1.63	0.85	0.98	0.95
Fumaric acid	0.90	0.77	0.67	nd
Vindoline	1.07	0.89	1.20	nd
Chlorogenic acid	1.54	0.85	0.75	nd
4-*O*-caffeoyl quinic acid	1.16	0.90	1.00	nd
Quercetin	1.50	1.33	nd	nd
Kaempferol	1.30	1.00	nd	nd
Secologanin	1.50	1.00	nd	nd
Loganic acid	1.10	1.00	0.94	0.89

nd, not detected.

Table 2. Comparison of metabolite levels in different organs between labeled and non-labeled *Catharanthus roseus* plants.

C-2 and C-3 imply that there might be a form of channeling that converts oxoglutarate C-4 to oxaloacetate C-2 or C-3 [18].

In plant cells, the labeling of amino acids alanine, glutamate, and aspartate is found to reflect that of the corresponding α-oxoacids: pyruvate, α-oxoglutarate, and oxaloacetate, respectively [19]. The organic acid malate showed a sixfold increased intensity for the carbon signal at δ 43.40.

Besides primary metabolites, secondary metabolites also exhibited clear ^{13}C incorporation. Two phenylpropanoids, chlorogenic acid and its isomer 4-*O*-caffeoyl quinic acid, have an increased ^{13}C intensity of C-6. Incorporation of ^{13}C could be observed for C-3 and C-10 of loganic acid. These results are in agreement with the prediction shown in **Figure 1**. The signal

Compound	Chemical Shift		Peak height		Relative intensity to CH_3OH-d_4		Relative enrichment ratio (*X*/*Y*)
	H	C	Control	Labeled	Control (*Y*)	Labeled (*X*)	
Glucose	4.58	97.04	3.2E+07	1.3E+08	0.17	2.89	17.31
Alanine	1.49	16.98	8.0E+06	2.1E+07	0.04	0.47	11.26
Glutamine	2.05	27.11	4.6E+07	3.9E+07	0.23	0.88	3.75
	2.38	31.83	1.6E+08	1.7E+08	0.80	3.72	4.63
	3.71	55.02	3.0E+08	2.3E+08	1.56	5.12	3.29
Glutamate	2.14	27.74	2.1E+06	7.0E+06	0.01	0.16	14.81
	2.46	34.44	6.3E+07	1.5E+08	0.32	3.36	10.45
	3.72	55.67	9.8E+06	3.9E+07	0.05	0.86	17.14
Arginine	1.72	24.9	1.7E+07	6.7E+06	0.09	0.15	1.77
	1.92	28.53	2.1E+07	8.1E+06	0.11	0.18	1.68
	3.24	41.38	5.9E+07	2.1E+07	0.30	0.47	1.56
Aspartate	2.64	37.21	1.0E+07	7.0E+06	0.05	0.16	3.05
Asparagine	2.83,2.96	35.23	1.2E+08	4.0E+07	0.62	0.90	1.44
	3.96	52.21	1.7E+08	4.7E+07	0.89	1.05	1.19
Threonine	1.34	20.47	8.8E+06	6.2E+06	0.05	0.14	3.08
Malate	2.35,2.72	43.4	6.0E+07	8.5E+07	0.31	1.91	6.21
Chlorogenic acid	7.07	123.12	1.2E+07	7.2E+06	0.06	0.16	2.53
4-*O*-Caffeoyl quinic acid	7.09	124.16	1.7E+06	2.5E+06	0.01	0.06	6.63
Loganic acid	1.07	12.69	2.5E+06	5.6E+06	0.01	0.13	9.96
	7.03	146.1	6.5E+05	5.2E+06	0.00	0.12	34.49
Vindoline	0.49	7.43	6.0E+06	2.4E+06	0.03	0.05	1.76

Table 3. The chemical shifts, peak height, and relative enrichment ratio of the same carbon signals in metabolites in labeled and non-labeled *Catharanthus roseus* plants.

corresponding to C-18 of vindoline in the labeled sample was twofold higher than in the spectrum of the control.

3.4. ^{13}C incorporation in different organs

Based on ^1H NMR spectra, relative levels of primary and secondary metabolites in different organs were calculated by normalizing the integral of signal peaks to the internal standard (TSP). **Table 4** showed that leaves, especially upper leaves, contained higher levels of amino acids, phenylpropanoids, iridoids, and vindoline, than stems and roots. In roots, phenylpropanoids and vindoline, which are biosynthesis dependent on chloroplasts, were not detected,

http://dx.doi.org/10.5772/65401

Compounds	Relative levels of metabolites			
	Upper leaf	Lower leaf	Stem	Root
β-Glucose	0.46	0.89	0.95	1.36
α-Glucose	0.34	0.56	0.62	0.83
Sucrose	0.20	0.33	0.53	0.81
Threonine	0.45	0.44	0.45	0.17
Alanine	0.45	0.46	0.29	0.17
Arginine	3.19	5.93	5.09	1.07
Glutamate	3.05	2.15	2.92	0.91
Glutamine	3.72	1.74	7.91	1.38
Aspartate	1.07	0.93	0.41	0.22
Asparagine	2.82	1.25	1.84	0.44
Malic acid	0.32	0.40	0.22	0.09
Fumaric acid	0.11	0.11	0.02	nd
Vindoline	0.28	0.09	0.05	nd
Chlorogenic acid	0.25	0.37	0.03	nd
4-*O*-caffeoyl quinic acid	0.17	0.15	0.04	nd
Quercetin	0.05	0.05	0.03	0.03
Kaempferol	0.09	0.10	0.02	nd
Secologanin	0.02	0.02	0.0006	0.0002
Loganic acid	0.07	0.14	0.11	0.05

Table 4. Relative level of metabolites in different organs of *Catharanthus roseus* based on ^{1}H NMR spectra.

whereas iridoids displayed a much lower level in roots while glucose and sucrose had relatively higher levels than in other organs.

The incorporation of ^{13}C in different organs (upper leaf, lower leaf, stem, and root) was also investigated by comparison of relative enrichment ratios in order to have a clue about the accumulation of label in different organs and its connection with transport and compartmentation of the pathways in the plants (**Table 5**). From the ^{13}C dimension of HSQC spectra of all organs, ^{13}C signals of labeled samples showed an apparently higher intensity in the amino acid and sugar areas than those of non-labeled ones (**Figure 6**), which indicated that ^{13}C-isotope was efficiently incorporated into the primary metabolism of intact *C. roseus* plants via the roots. Glucose had higher ^{13}C intensity ratio in lower leaves and roots but relatively low in upper leaves and stems, thus showing a time-dependent distribution through the plant. Glutamate and aspartate, directly derived from α-ketoglutarate and oxaloacetate of the TCA cycle, showed clear ^{13}C enrichment in all organs. So did malate, one of the bricks in the TCA cycle. Meanwhile, glutamate, aspartate, and malate all displayed

Compounds	^{13}C Chemical shift	Relative enrichment ratio (labeled:control)			
		Upper leaf	Lower leaf	Stem	Root
Alanine	16.98	9.76	32.38	118.89	84.86
Threonine	20.47	3.55	1.06	1.39	18.91
Arginine	24.9	1.89	1.12	1.67	2.99
	28.53	1.58	1.07	1.17	1.15
	41.38	1.52	0.91	0.91	1.66
Glutamine	27.11	2.80	2.34	1.89	6.38
	31.83	3.21	4.03	3.09	6.70
	55.02	2.34	1.38	2.01	8.43
Glutamate	27.74	11.79	5.39	3.36	20.65
	34.44	7.21	3.37	4.63	16.75
	55.67	15.49	2.72	5.11	25.06
Asparagine	35.23	1.21	0.75	1.09	5.67
	52.21	1.17	0.95	1.23	4.18
Aspartate	37.21	2.25	3.40	4.07	36.29
Malate	43.4	4.67	4.41	7.82	26.51
β-glc	97.04	30.96	55.55	15.01	32.79
Vindoline	7.43	2.96	nd	nd	nd
Loganic acid	12.69	7.66	3.75	4.26	23.62
	146.1	27.96	13.77	6.59	24.79
Chlorogenic acid	123.12	2.88	1.42	nd	nd
	146.8	93.69	nd	nd	nd
4-*O*-Caffeoyl quinic acid	124.16	10.35	nd	nd	nd

nd, not detected.

Table 5. Relative enrichment ratios of the carbons of some metabolites in different organs of *Catharanthus roseus* plants fed with [1-^{13}C] glucose.

the highest ^{13}C intensity ratio in roots. These results indicate that ^{13}C was efficiently incorporated and recycled in the primary metabolism of intact plants. Upper leaves had higher levels and higher relative enrichment ratios of glutamate and malate compared with lower leaves, reflecting the faster rate of TCA cycle in the upper parts for plants growing. The glutamate-derived amino acids glutamine and arginine displayed a different pattern of ^{13}C incorporation. Glutamine showed ^{13}C incorporation in all organs with the highest intensity ratio in roots and the lowest in stems, whereas arginine showed low ^{13}C incorporation in all organs, implying a low flux in its biosynthetic pathway and low usage for other pathways. The ^{13}C incorporation of aspartate-derived amino acids asparagine and threonine was also

http://dx.doi.org/10.5772/65401

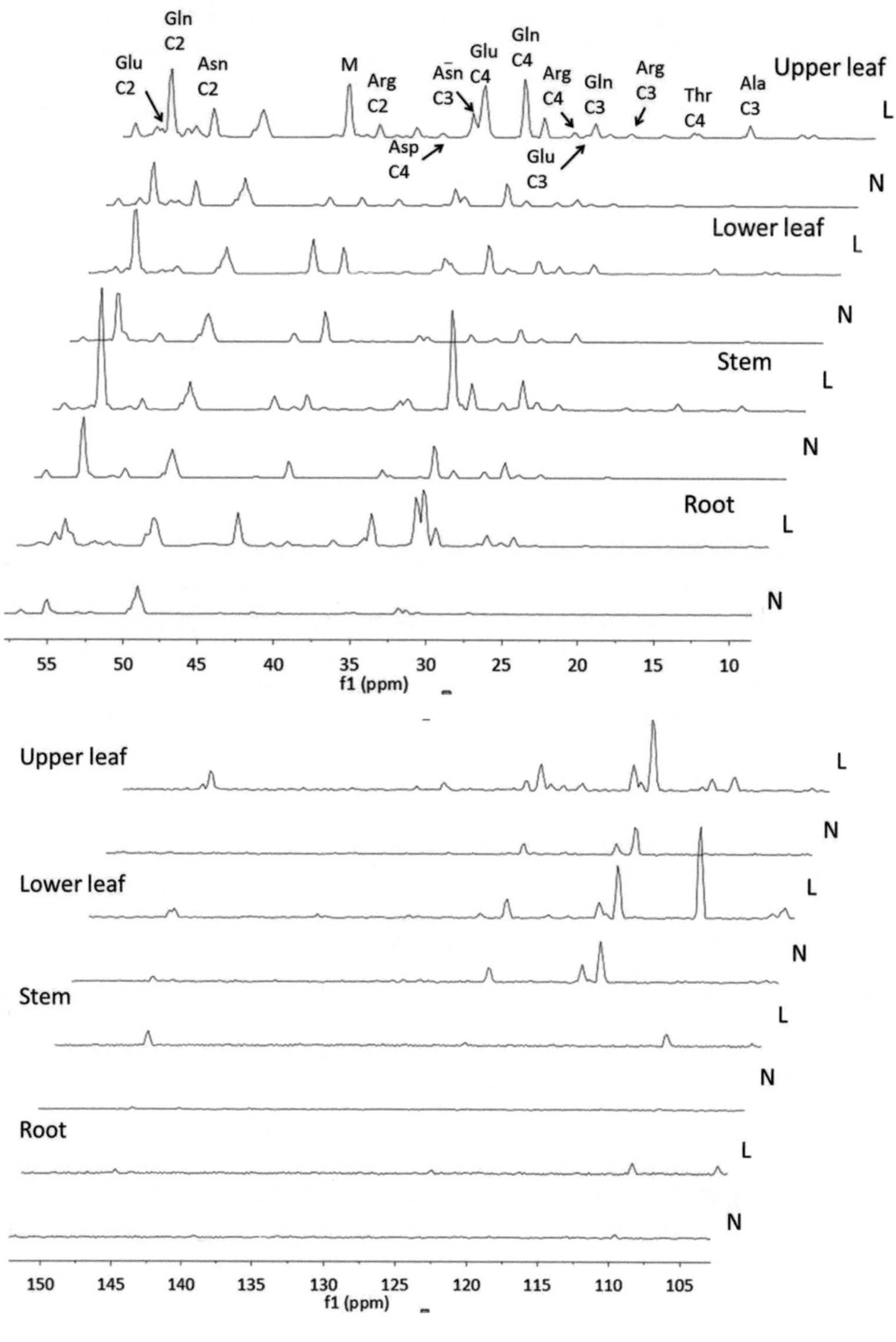

Figure 6. ^{13}C dimension of HSQC spectra of amino acids (δ 10–55 ppm) and secondary metabolites (δ 105–150 ppm) in different organs of *Catharanthus roseus* after feeding [1-^{13}C] glucose. L, labeled samples; N, non-labeled samples; M, malate.

different. Threonine had relatively high ^{13}C incorporation in upper leaves and roots, but relatively low in lower leaves and stems, indicating a high turnover in the upper leaves. Asparagine, except for roots, displayed low ^{13}C incorporation in the other organs. Pyruvate-derived alanine exhibited the lowest relative enrichment ratio in upper leaves, while the highest was found in stems and roots.

Based on the ^{13}C dimension of HSQC spectra, leaves had more ^{13}C signals in the area of >δ 100 ppm than stems and roots (**Figure 6**), even after feeding [1-^{13}C] glucose. Upper leaves had relatively high ^{13}C incorporation for vindoline, chlorogenic acid, and 4-*O*-caffeoyl quinic acid while lower leaves only showed ^{13}C incorporation into chlorogenic acid. The levels of these phenylpropanoids in lower leaves were almost as much as in upper leaves. This means that all three compounds have a high biosynthetic rate in the upper leaves, whereas in the lower leaves the biosynthesis of chlorogenic acid is more active than of the other two compounds. In plants, lower leaves, with older age, are entering the senescence phase, among others reflected by a lower rate of both primary and secondary metabolism. Studies in previous chapters have shown that vindoline levels show an age-related decrease, consistent with previously reported results [20].

In stems and roots, no ^{13}C signals of vindoline, chlorogenic acid, and 4-*O*-caffeoyl quinic acid were detected with or without feeding [1-^{13}C] glucose. Vindoline is not found in roots due to its tissue-specific biosynthesis requiring chloroplasts for one of its biosynthetic steps [20–25]. The ^{13}C signal of loganic acid at δ 12.69 and 146.1 ppm was clearly present and showed a high relative enrichment ratio at the spectra of all organs while that of secologanin at δ 121.53 ppm was only found in the spectra of leaves. It was difficult to calculate the relative enrichment ratio of secologanin due to the signal overlapping. In roots and stems, secologanin was too low for further analysis. The high levels of loganic acid in the roots are in line with a previous study that reported that LAMT activity, which converts loganic acid into loganin (the direct substrate of secologanin), was four to eight times lower in hairy roots than that in the other organs of the plant [22].

3.5. Effect of JA elicitation on ^{13}C fluxes into metabolic pathways

JA was spiked into the labeled glucose solution at the sixth day after submerging the plant roots in the solution. The control plants were also reared in labeled glucose solution but without JA elicitation. Leaves were harvested at 0, 6, 24, and 72 h (6, 7, and 9 d of incubation with the labeled glucose solution) after elicitation and measured by ^{1}H NMR and HSQC.

For control plants, NMR spectra showed that the enrichments of malic acid and of the amino acids alanine, arginine, glutamate, glutamine, aspartate, and asparagine in the leaves were nearly identical at 6 and 9 d of incubation with the labeled glucose solution (**Figure 7**), suggesting the establishment of steady state at 6 d. However, the incorporation of label in glucose and threonine increased continuously within the measured period of 9 days. Besides, loganic acid and chlorogenic acid kept the same enrichments while vindoline and 4-*O*-caffeoyl quinic acid showed an increase of the enrichments within 9 days. Previous study with *C. roseus* hairy roots grown in the light showed that the ^{13}C label was not diluted by CO_2 fixation [7]. In tobacco plants grown on agar containing labeled glu-

http://dx.doi.org/10.5772/65401

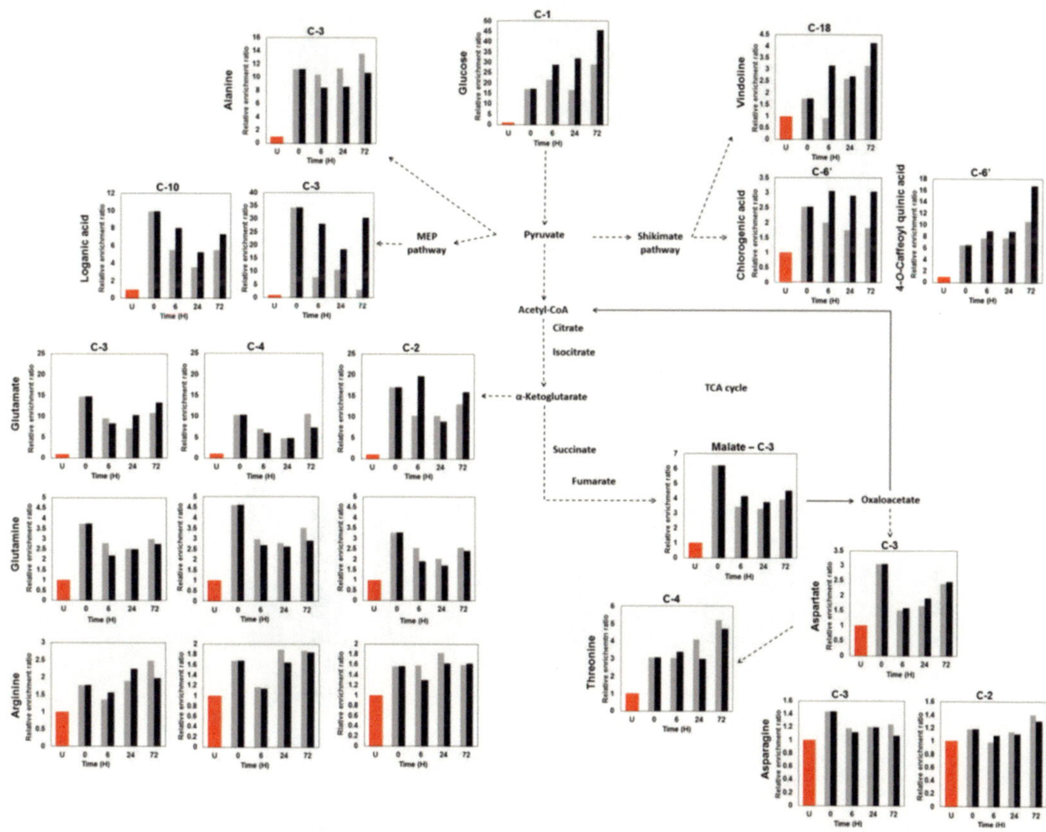

Figure 7. Relative enrichment ratio of primary and secondary metabolites during incubation of *Catharanthus roseus* plants with [1-^{13}C] glucose. Gray bars: JA-elicited samples; black bars: control samples; red bars (U): unlabeled samples (without incubation in [1-^{13}C] glucose solution).

cose, the metabolism was studied on a quantitative basis showing that the labeled glucose was efficiently absorbed via the root system, metabolized, and recycled [26]. Our results indicate that the *C. roseus* plant system can reach a relatively steady isotopic state with plants growing in ^{13}CLE.

JA elicitation had little effect on the level of most metabolites, except glutamate, glutamine, vindoline, and loganic acid. Although JA induced an increase of glutamate and glutamine levels (**Figure 8**), their relative enrichment ratio remained unchanged compared with the controls. At the same time, the enrichment of alanine at C-3 showed an increase without levels changing compared to the controls. Vindoline levels showed an increase and reached the highest level at 72 h (23% higher than the controls) after JA treatment (**Figure 8**). However, the relative enrichment ratio of the C-18 signal of vindoline was lower in JA-elicited samples than in the controls, especially at 6 h (**Figure 7**). The level of loganic acid decreased with time (**Figure 8**), leading to a dramatical decrease of its enrichment at both C-3 and C-10 from 6 to 72 h. The levels of chlorogenic acid and 4-*O*-caffeoyl

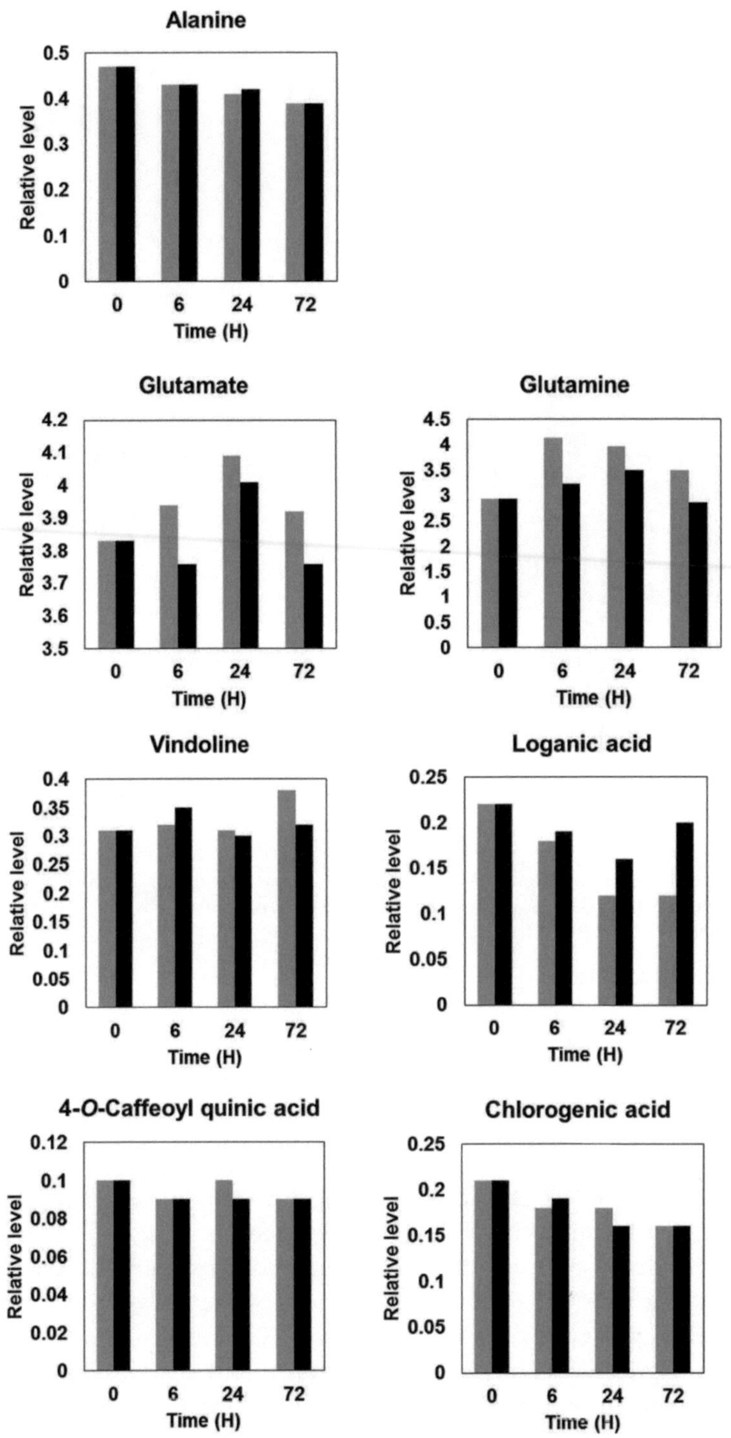

Figure 8. Relative levels of metabolites in *Catharanthus roseus* leaves after JA elicitation. Gray bars: JA-elicited samples; black bars: control samples.

http://dx.doi.org/10.5772/65401

quinic acid in the time course did not change after JA elicitation (**Figure 8**), but the enrichments were lower than those of the control-labeled samples (**Figure 7**). ^{13}C fluxes to various metabolic pathways, such as glutamate and loganic acid, could be disturbed within 24 h after MeJA treatment [27].

4. Future prospects

Metabolic flux analysis is the quantification of all intracellular fluxes in an organism, which is thus an important cornerstone of metabolic engineering and systems biology. This study reports a comprehensive ^{13}C-labeling-based metabolomics of a plant system. [1-^{13}C] glucose was efficiently absorbed via the root system and recycled in the whole plant of *C. roseus*. The plant system of *C. roseus* could reach a relatively steady isotopic state in ^{13}CLE, which appears to be well qualified to study flux contributions in the biosynthesis of sink metabolites for system biology. Combined with exogenous elicitation, ^{13}C MFA appears to be a good tool to study the crosslink among pathways in the complicated plant metabolic network. The development of a comprehensive flux analysis tool for the plant system of *C. roseus* is expected to be valuable in assessing the metabolic impact of genetic or environmental changes.

Author details

Qifang Pan[1*], Natali Rianika Mustafa[2], Robert Verpoorte[2] and Kexuan Tang[1]

*Address all correspondence to: panqf@sjtu.edu.cn

1 Plant Biotechnology Research Center, SJTU-Cornell Institute of Sustainable Agriculture and Biotechnology, Fudan-SJTU-Nottingham Plant Biotechnology R&D Center, School of Agriculture and Biology, Shanghai Jiaotong University, Shanghai, PR China

2 Natural Products Laboratory, Institute of Biology, Leiden University, Leiden, The Netherlands

References

[1] Ratcliffe RG, Shachar-Hill Y. Measuring multiple fluxes through plant metabolic networks. Plant J. 2006;45: 490–511.

[2] Szyperski T. C-NMR, MS and metabolic flux balancing in biotechnology research. Q Rev Biophys. 1998;31: 41–31106.

[3] Möllney M, Wiechert W, Kownatzki D, de Graaf AA. Bidirectional reaction steps in metabolic networks: IV. Optimal design of isotopomer labeling experiments. Biotechnol Bioeng. 1999;66: 86–103.

[4] Contin A, van der Heijden R, Lefeber AWM, Verpoorte R. The iridoid glucoside secologanin is derived from the novel triose phosphate/pyruvate pathway in a *Catharanthus roseus* cell culture. FEBS Lett. 1998;434: 413–416.

[5] Mustafa NR, Kim HK, Choi YH, Erkelens C, Lefeber AW, et al. Biosynthesis of salicylic acid in fungus elicited *Catharanthus roseus* cells. Phytochemistry. 2009;70: 532–539.

[6] Sriram G, Fulton DB, Shanks JV. Flux quantification in central carbon metabolism of *Catharanthus roseus* hairy roots by 13C labeling and comprehensive bondomer balancing. Phytochemistry. 2007;68: 2243–2257.

[7] Schuhr CA, Radykewicz T, Sagner S, Latzel C, Zenk MH, et al. Quantitative assessment of crosstalk between the two isoprenoid biosynthesis pathways in plants by NMR spectroscopy. Phytochem Rev. 2003;2: 3–16.

[8] Lundström P, Teilum K, Carstensen T, Bezsonova I, Wiesner S, et al. Fractional ^{13}C enrichment of isolated carbons using [1-^{13}C]-or [2-^{13}C]-glucose facilitates the accurate measurement of dynamics at backbone Cα and side-chain methyl positions in proteins. J Biomol NMR. 2007;38: 199–212.

[9] Murashige T, Skoog F. A revised medium for rapid growth and bio assays with tobacco tissue cultures. Physiol Plantarum. 1962;15: 473–497.

[10] Kim HK, Khan S, Wilson EG, Kricun SDP, Meissner A, et al. Metabolic classification of South American Ilex species by NMR-based metabolomics. Phytochemistry. 2010;71: 773–784.

[11] Meyer RC, Steinfath M, Lisec J, Becher M, Witucka-Wall H, et al. The metabolic signature related to high plant growth rate in *Arabidopsis thaliana*. Proc Nat Acad Sci. 2007;104: 4759–4764.

[12] Lea PJ, Sodek L, Parry MA, Shewry PR, Halford NG. Asparagine in plants. Ann App Biol. 2007;150: 1–26.

[13] Genix P, Bligny R, Martin J-B, Douce R. Transient accumulation of asparagine in sycamore cells after a long period of sucrose starvation. Plant Physiol. 1990;94: 717–722.

[14] Choi YH, Tapias EC, Kim HK, Lefeber AWM, Erkelens C, et al. Metabolic discrimination of *Catharanthus roseus* leaves infected by phytoplasma using ^{1}H-NMR spectroscopy and multivariate data analysis. Plant Physiol. 2004;135: 2398–2410.

[15] Mustafa NR, Kim HK, Choi YH, Verpoorte R. Metabolic changes of salicylic acid-elicited *Catharanthus roseus* cell suspension cultures monitored by NMR-based metabolomics. Biotechnol Let. 2009;31: 1967–1974.

[16] Kruger NJ, Huddleston JE, Le Lay P, Brown ND, Ratcliffe RG. Network flux analysis: Impact of 13 C-substrates on metabolism in *Arabidopsis thaliana* cell suspension cultures. Phytochemistry. 2007;68: 2176–2188.

[17] Malloy CR, Sherry AD, Jeffrey F. Evaluation of carbon flux and substrate selection through alternate pathways involving the citric acid cycle of the heart by 13C NMR spectroscopy. J Biol Chem. 1988;263: 6964–6971.

[18] Dieuaide-Noubhani M, Raffard G, Canioni P, Pradet A, Raymond P. Quantification of compartmented metabolic fluxes in maize root tips using isotope distribution from 13C-or 14C-labeled glucose. J Biol Chem. 1995;270: 13147–13159.

[19] Salon C, Raymond P, Pradet A. Quantification of carbon fluxes through the tricarboxylic acid cycle in early germinating lettuce embryos. J Biol Chem. 1988;263: 12278–12287.

[20] Shukla AK, Shasany AK, Gupta MM, Khanuja SP. Transcriptome analysis in *Catharanthus roseus* leaves and roots for comparative terpenoid indole alkaloid profiles. J Exp Bot. 2006;57: 3921–3932.

[21] Zhou ML, Zhu XM, Shao JR, Tang YX, Wu YM. Production and metabolic engineering of bioactive substances in plant hairy root culture. Appl Microbiol Biotechnol. 2011;90: 1229–1239.

[22] Murata J, Roepke J, Gordon H, De-Luca V. The leaf epidermome of *Catharanthus roseus* reveals its biochemical specialization. Plant Cell. 2008;20: 524–542.

[23] Abbasi BH, Tian CL, Murch SJ, Saxena PK, Liu CZ. Light-enhanced caffeic acid derivatives biosynthesis in hairy root cultures of *Echinacea purpurea*. Plant Cell Rep. 2007;26: 1367–1372.

[24] Murata J, De-Luca V. Localization of tabersonine 16-hydroxylase and 16-OH tabersonine-16-O-methyltransferase to leaf epidermal cells defines them as a major site of precursor biosynthesis in the vindoline pathway in *Catharanthus roseus*. Plant J. 2005;44: 581–594.

[25] De Luca V, Cutler AJ. Subcellular localization of enzymes involved in indole alkaloid biosynthesis in *Catharanthus roseus*. Plant Physiol. 1987;85: 1099–1102.

[26] Ettenhuber C, Radykewicz T, Kofer W, Koop H-U, Bacher A, et al. Metabolic flux analysis in complex isotopolog space. Recycling of glucose in tobacco plants. Phytochemistry. 2005;66: 323–335.

[27] Antonio C, Mustafa NR, Osorio S, Tohge T, Giavalisco P, Willmitzer L, Rischer H, Oksman-Caldentey KM, Verpoorte R, Fernie AR. Analysis of the interface between primary and secondary metabolism in *Catharanthus roseus* cell cultures using 13C-stable isotope feeding and coupled mass spectrometry. Mol Plant. 2013;6: 581–584.

Chapter 6

Metabolomics Approaches and their Hidden Potential for Explaining the Mycotoxin Contamination Problem

Laura Righetti, Chiara Dall'Asta, Jana Hajslova and
Josep Rubert

Additional information is available at the end of the chapter

http://dx.doi.org/10.5772/65647

Abstract

Food is essential for life. On the basis of the previous sentence, consumers have a right to expect that the foods they purchase and consume will be safe, authentic and of high quality. On these premises, target compounds, such as mycotoxins, pesticides or antibiotics, have been commonly investigated on the food chain, and subsequently, were regulated by authorities. This raises the following question: may consumer be prevented to these risk exposures? Probably not, food chain is step-by-step longer and more complex than ever before. Note that food chain is affected by globalized trade, culture, travel and migration, an ageing population, changing consumer trends and habits, new technologies, emergencies, climate change and extreme weather events which are increasing foodborne health risks, especially for mycotoxins. Because of the fact that mycotoxins are natural toxic compounds produced by certain filamentous fungi on many agricultural communities. In fact, these toxins have adverse effects on humans, animals and crops that result in illnesses and economic losses. Nevertheless, so far mycotoxins and their modified forms have been mainly monitored in cereal and cereal-based products, however, may an early detection of mycotoxins be considered a reliable strategy? In this chapter, recent metabolomics approaches have been reviewed in order to answer this question and to understand future strategies in the field of mycotoxin contamination.

Keywords: food metabolomics, mycotoxins, plant metabolome, fungal pathogens

1. Introduction

Mycotoxins are secondary metabolites (300–800 Da) produced by filamentous fungi that colonize crops in field and upon storage, being among them cereals one of the most affected commodities [1]. Fungal colonization is strongly dependent on environmental conditions and agricultural practices. Climatic factors such as temperature, humidity, rainfalls, as well as the concomitant presence of other pests or insects may support the fungal infection. Therefore, climate change is significantly affecting the mycotoxin contamination of crops worldwide. As a consequence, fungal infection and related pathogenic diseases can cause significant yield losses, quality reductions and mycotoxins accumulation in crops, particularly grains [2].

Although regulations, adequate quality controls and good agricultural practices have been implemented in many countries, the mycotoxin contamination represents a serious challenge for global trade in terms of animal and human health threat and economical losses. For this reason, the establishment of common standard procedure for fungal biocontrol and mycotoxin mitigation are under investigation.

From a toxicological perspective, mycotoxins can cause both acute and chronic effects for humans and animals. They are responsible for a broad spectrum of toxic activities, ranging from severe adverse effects on the liver, kidney, hematopoetic, immune system, foetal and reproductive systems, as well as significant contribution to carcinogenic and mutagenic developments [3]. In fact, The International Agency for Cancer Research (IARC) has formally classified a number of mycotoxins. For example, four aflatoxins are classified in Group 1 (AFB_1, AFB_2, AFG_1 and AFG_2) while ochratoxin A (OTA) is classified in Group 2B [4, 5].

Among them, those produced by *Fusarium* spp. are often found in cereals, and are related to pathogenic diseases in plants, as well. In particular, *Fusarium* Head Blight (FHB) is recognized as one of the most destructive global diseases of wheat and barley [6]. FHB can cause, indeed, significant reductions in grain yield and quality, and is associated with the accumulation of mycotoxins, such as deoxynivalenol (DON). Thus, besides the severe economic impact, due to losses in productivity, FHB represents a serious health risk for consumers and livestock [3]. In order to reduce the economic and health impact of FHB, several cultural practices have been proposed so far. However, crop rotation, tillage, use of fungicides or other biocontrol agents are generally regarded as insufficient to tackle FHB and mycotoxins contamination alone [3]. This is mainly due to the fact that the breeding of grains for superior technological properties has led to a decrease of the genetic diversity, with a subsequent increase of susceptibility towards pathogenic diseases. Therefore, the study of the plant response to fungal infection is crucial for developing possible strategies to counteract mycotoxin accumulation.

From a biological point of view, the role of mycotoxins in fungal colonization is still to be clarified. Some of them—such as deoxynivalenol (DON) have been proved to be virulence factor for fungal infection [7]. However, the intense cross-talk among plant and pathogen affects the biological cascade, from genes to metabolites, and plays a significant role in mycotoxin accumulation. Fungal infection and mycotoxin contamination are commonly addressed with classical methods, from DNA-based techniques for fungal identification to analytical methodologies for mycotoxin detection. The residual DNA content of fungal pathogens was used to identify unequivocally fungal species, and they were associated with cereals and their mycotoxins [8], basically allowing for a toxigenic fungi monitoring. However, the main disadvantage of this technique is associated to relatively high cost and the fact that it is time-consuming. In addition, a poor correlation between fungal growth and mycotoxin accumulation has been pointed out.

For this reason, classical chromatographic methods are often used for mycotoxin determination in crops and products thereof [9–11]. Over the last decade, mass spectrometry (MS)-based methods have become the golden standard for mycotoxin analysis, being the multitoxin approach the most promising strategy to control the occurrence of multiple analytes in the same material [12]. As a complementation, rapid diagnostic methods are commonly based on immunochemical assays (i.e., lateral flow devices, dipsticks, etc.) for early detection at pre- and post-harvest [13]. More recently, nondestructive imaging methods have been proposed as well as rapid diagnostic tool [14]. In this context, the untargeted methodologies have started to be applied only recently, and only to meet specific needs. In particular, the omics strategies have been applied to the mycotoxin issue to investigate the interaction between the plant and the pathogen in field, leading to mycotoxin accumulation [3, 15–19].

In a top-down view, genomics and transcriptomics studies have proposed to investigate the biosynthetic pathways for mycotoxin production, and their regulation upon biotic and abiotic stress. Similarly, proteomics has been often proposed for identifying enzymes and proteins responsive to pathogenic diseases, such as FHB [20], or responsible for mycotoxin modification in plant [21, 22]. Over the last decade, however, the field of metabolomics has gained increasing interest across all disciplines, and has found a prominent role in mycotoxin-related studies as well. Metabolomics is an emerging technique that can be considered complementary to the other omics approaches and highlighting unique advantages. A metabolic fingerprint may generate thousands of data points, of which only a handful might be needed to describe the problem adequately [23, 24]. Extracting the most meaningful elements of these data is thus key to generating useful new knowledge with mechanistic or explanatory power.

To date, however, in the vast majority of cases, mycotoxin contamination has been directionally explored. In this way, up to now, the mycotoxin contamination loop has not been properly closed and many issues are still open. One of the main challenges in mycotoxin analysis will be to improve our limited understanding of the roles of plant pathogen cross-talk at the molecular level. In this context, a multiomics global strategy may be able to identify chemical markers at the earliest stage, and to univocally characterize resistant varieties and the early detection of mycotoxins. The early detection of toxigenic fungi or of markers of the interaction between the pathogen and its host can be usefully exploited to limit entering of mycotoxins into the food/feed production chain.

2. Advanced analytical tools merged with chemometrics

The multiomics approach has been poorly compared to classical approaches described in the previous section. Initially, innovative spectral techniques (i.e., imaging analysis, near-infrared, Raman) have been proposed for the early detection of fungal pathogens [25, 26]. Since fungal growth is not strictly related to mycotoxin accumulation, and to the pattern of occurring mycotoxins, these techniques cannot provide a response on mycotoxin occurrence or chemical markers, mainly linked to the plant-pathogen interactions. In this framework, metabolomics may represent the golden tool for understanding the biological pathways involved in mechanisms of plant resistance. Nowadays, gas chromatography (GC) and liquid chromatography (LC) are commonly used for metabolomics approaches, mainly coupled to mass spectrometry (MS) [3]. In principle, LC-MS and GC-MS provide a high number of scans per peak, allowing peak picking and alignment (feature extraction), and if necessary quantification, as well as a large dynamic range in order to monitor low and high concentration levels of metabolites.

2.1. Liquid chromatography coupled to mass spectrometry (LC-MS)

LC-MS has been the most commonly used metabolic fingerprinting/profiling approach for understanding plant resistance mechanisms and the plant-pathogen cross-talk. For instance, Cajka et al. [27] have recently developed an analytical procedure based on the optimization of a solid-liquid extraction procedure using methanol/water (50:50, v/v), in order to isolate polar/medium-polar barley metabolites followed by ultra high performance liquid chromatography quadrupole-time-of-flight (UHPLC-QTOF) [27]. **Figure 1** shows unique and shared metabolites acquired by UHPLC-QTOF using both positive and negative ionization modes.

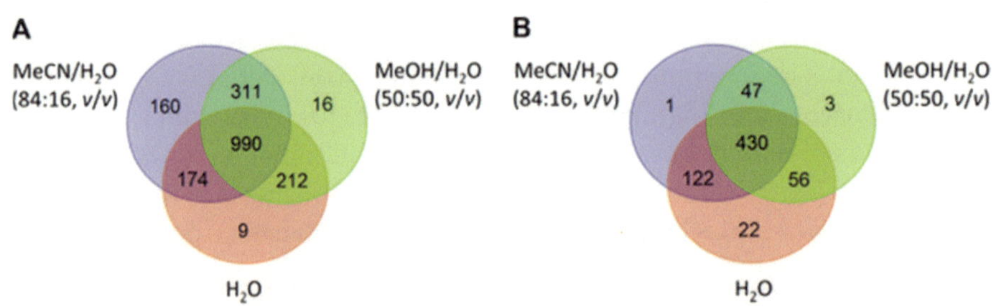

Figure 1. Venn diagrams illustrating shared and unique features in barley extracts prepared under the different extraction procedures and analyzed by both positive (A) and negative (B) ionization modes UHPLC-QTOF.

The authors demonstrated how the carefully in-depth investigation of sample preparation could support the extraction of the broadest spectrum of metabolites isolated from the matrix, in this particular case barley. Obviously, UHPLC-QTOF chemical fingerprints differed significantly depending on the extraction solvent used (see **Figure 2**). For example, when deionized water was used, a lower extraction efficiency of less polar compounds was exhibited. Nevertheless, sample preparation using a mixture of acetonitrile/water (84:16, v/v) or methanol/water (50:50, v/v) enhanced the extraction of less polar and polar compounds were also detected. The authors, as a compromise, chose methanol/water (50:50, v/v), since the extraction mixture permitted isolation of both highly polar and less polar metabolites. So far, various proportion of aqueous methanol has been mainly applied, as it can be seen in **Table 1**. In this way, the changes occurring both in primary carbohydrates and primary nitrogen metabolism upon plant infection have been partially elucidated. On the other hand, lipidomic approaches applying more nonpolar solvent (e.g., hexane, dichloromethane, ethyl acetate) have been exclusively used to investigate the plant-pathogen cross-talk in maize [28–30]. Increasing evidence indicates, indeed, that lipid signalling is an integrated part of the complex regulatory network in plant pathogen cross-talk.

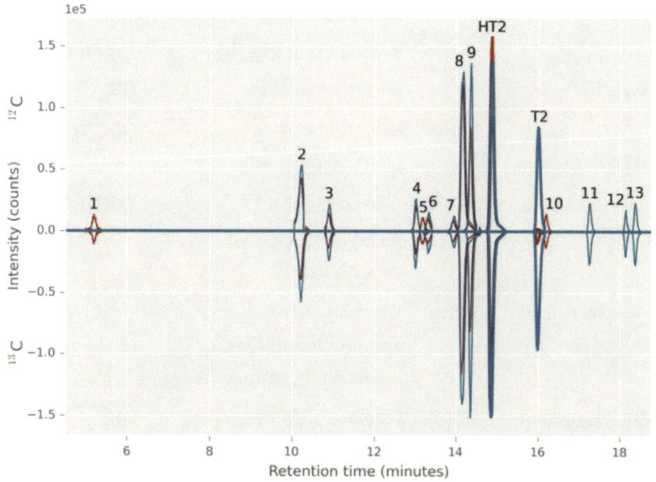

Figure 2. Overlaid extracted ion chromatograms (EICs) based on MetExtract data processing output showing the biotransformation products of a sample treated with a mixture of $^{12}C/^{13}C$-HT-2 toxin (red trace) and one treated with a mixture of $^{12}C/^{13}C$ T-2 toxin (blue trace). EICs of nonlabeled metabolites were displayed with positive intensities; those of the corresponding labeled metabolites were displayed as negative intensities.

Not only fingerprinting approaches, but also metabolic profiling strategy has been recently performed using a stable isotopic labelling approach in order to understand the metabolic fate of HT-2 toxin and T-2 toxin in wheat [31]. In general, untargeted metabolomics approaches are usually based on generic settings for sample preparation (which usually include a simple extraction without any purification step, or nonsample preparation), separation and detection. In contrast, if a particular group of metabolites is preselected, a metabolic profiling is carried out. Thereby, a more specific extraction procedure and chromatographic separation/detection has to be performed. In this way, this study was focused on Type A trichothecenes, such as HT2 and T2 toxins, and their detoxification pathways.

The stable isotopic labelling approach applied is really innovative since monitoring pairs of corresponding nonlabeled and labeled precursor allowed metabolome to be easily monitored and interpreted, providing further information. Liquid chromatography high-resolution tandem mass spectrometry (LC-HRMS/MS) spectra of the observed metabolites of HT2 and T2 were compared with those obtained in wheat and were shown to be identical. **Figure 2** shows overlaid extracted ion chromatograms of all detected biotransformation products. In this frame, the authors demonstrated that the exposure of wheat to either HT2 or T2 toxins primarily activates biotransformations involving hydroxylation, (de)acetylation and various conjugations. Furthermore, kinetic data revealed that detoxification progressed rapidly, resulting in the almost complete degradation of the toxins, within 1 week, after a single exposure.

Instrumentation	Extraction	Plant	Chemical group	Markers class	References
LC-HRMS	Water/methanol (45:65, v/v)	Barley	Fatty acids, flavonoid phenylpropanoids, amino acids, terpenoids, organic acids	RRC, PRr, PRs, RI	Bollina et al. [16]
	Water/methanol (50:50, v/v)	Barley	Fatty acids, flavonoid phenylpropanoids	RRI, RRC, RI	Bollina et al. [17]
	Water/methanol (50:50, v/v)	Barley	Fatty acids, phenylpropanoids	RR, RI	Cajka et al. [27]
	Water/methanol (40:60, v/v)	Wheat	Fatty acids, phenylpropanoids, terpenoids	RRI, RRC, RI, PRp	Gunnaiah et al. [18]
	Water/methanol (40:60, v/v)	Wheat	Flavonoid phenylpropanoids, terpenoids, amino acids, carbohydrates	RRC, RRI, RI	Gunnaiah et al. [19]
GC-MS	Water/methanol /formic acid (74:25:1, v/v)	Wheat	Polyamines, amino acids, phenylpropanoids, carbohydrates	RR	Warth et al. [15]
	Water/methanol /formic acid (74:25:1, v/v)	Wheat	Amino acids, amines, carbohydrates	RR	Nussbaumer et al. [36]
	Water/methanol (50:50, v/v)	Wheat	Polyamines, amino acids, phenylpropanoids, carbohydrates	RR	Paranidharan et al. [35]
H NMR	Methanol/water (40:60, v/v)	Wheat	Amines, amino acids, carbohydrates	RR, PR	Browne et al. [33]
	Methanol/water (40:60, v/v)	Wheat	Amines, amino acids, carbohydrates, phenylpropanoids	RR, PR	Cuperlovic-Culf et al. [24]

RRC, resistance-related constitutive; RRI, resistance-related induced; RI, resistance indicator; PR, pathogenesis-related; PRr, pathogenesis-related resistant; PRs, pathogenesis-related susceptible; PRp, pathogenesis-related proteins.

Table 1. Putative metabolites involved in *Fusarium* Head Blight resistance, reported in the literature so far.

2.2. Gas chromatography-mass spectrometry (GC-MS)

Surprisingly, GC coupled to high-resolution mass spectrometry (HRMS) has not been applied to metabolomics strategies. As it was discussed above, the applicability of HRMS permitted metabolic pathways to be clearly described. Nevertheless, GC coupled to a single quadrupole shows selectivity and specificity for metabolomics approaches, since available databases containing mass spectra and retention indexes can tentatively identify metabolites due to the extensive and reproducible fragmentation pattern obtained in full-scan mode using electron ionisation (EI). A recent research work was focused on the applicability of GC-EI-MS in order to understand deoxynivalenol (DON) accumulation in wheat [15]. In this research, the experimental design was nicely described, and similar to previous research described above. Nevertheless, sample preparation took extra time compared to LC-MS, due to derivatization procedure based on silylation. Many metabolites contain polar functional groups and are thermally labile for separation by GC or present limited volatility, therefore, derivatization often has to be applied. Oximation or silylation has been commonly applied due to their universality and versatility [24].

2.3. Data processing to extract meaningful markers

For processing massive information based on separation techniques and mass spectrometry, effective software tools capable of rapid data mining procedures have to be used. Note that data matrices contain thousands of variables (*m/z*, RT, intensity), and they have to be converted into more manageable information [24].

Data processing and data pretreatment must be carried out in order to permit the identification of significant metabolites, which capture the bulk of variation between different datasets and may therefore potentially serve as biomarkers. Data processing usually involves four basic steps: deconvolution, alignment, filtering and gap filling. The features, defined by their *m/z* and retention time, and their intensities in different samples are used for the statistical analysis. Samples would be grouped and it can be observed using scores plots, heatmaps or hierarchical clustering. After data pretreatment, a statistical comparison can be performed using the multivariate data analysis (MVDA). Usually this step involves unsupervised models (PCA) and supervised classification tools, such as PLS-DA and OPLS-DA. These supervised methods are performed to maximize differences between groups and to highlight potential biomarkers. When the experimental design is more complex, the use of *t*-test or other univariate data analysis (UVDA) tools represents the best choice [32].

3. Metabolomics to decipher pathways involved in *Fusarium* Head Blight resistance

As it was already mentioned in the Introduction section, mycotoxins and fungal pathogens, such as *Fusarium graminearum*, can cause global diseases of wheat (*Triticum aestivum* L.) and barley [6]. Nevertheless, up to now, all preventive techniques used have been pointless, such as fungicides or crop rotation. Breeding strategies for increasing pathogen resistance seem to be the most promising and environmentally safe strategy for controlling mycotoxin accumulation in grains. It is known, indeed, that plant resistance mechanisms may be controlled by several quantitative trait loci (QTLs) that contribute to overall pathogen resistance in three different ways classified as type 1, 2 and 3, and referred as resistance to initial infection of spikelets, spread of pathogen within spikes and accumulation of mycotoxins, respectively. The involved QTLs typically are linked to, or contain, the genes that control the phenotype. Over hundred of QTLs for FHB resistance in wheat have been already identified [3, 7, 15, 32]. However, fully resistant varieties are still to be identified or inbred. Thus, there is an urgent need to better understand the mechanisms of resistance against *Fusarium* spp. in order to develop novel strategies and resistance varieties.

Nowadays, recent advances in metabolomics offer new opportunities to elucidate complex metabolic pathways involved in *Fusarium* resistance and potential FHB resistance biomarker metabolites in barley and wheat [3, 15–19, 32]. In fact, during the last decade, the applicability of metabolomics has significantly increased in this field. Nevertheless, knowledge remains still partial, and a long way has to be covered towards the development and understanding of the plant-pathogen interactions. This new scenario will provide suitable knowledge related to plant metabolome, which was already explained by a few examples in the previous section.

Different strategies have been applied so far, NMR for polar metabolites [33, 34], LC-QTOF for semipolar metabolites [16–19, 32] and GC-MS for volatile compounds [15, 35, 36]. However, we should keep in mind that a strategy able to simultaneously extract and detect the entire metabolome does not exist. Consequently, the data delivered by metabolomics studies only cover a fraction of the metabolome. In other words, the picture taken exclusively reveals one part of the metabolome. In addition, the resistance mechanism is a result of multi-interactions between biomolecules such as genes, proteins, metabolites and environmental factors. Therefore, a multiomics approach based on proteomics and metabolomics could overcome any limitation in the experimental design. For example, an integrated nontargeted metabolo-proteomics approach was recently published [18, 32]. This strategy demonstrated to be a powerful tool for a more comprehensive analysis in order to elucidate the mechanism, revealing successfully changes in the wheat primary metabolism, in response to *F. graminearum*.

4. Setting up of the experimental plan

Depending on the hypotheses to be tested, different combinations of plants and fungal pathogens can be employed to explore the system relationship. Up to date, the metabolomic approaches have been mainly restricted to study the resistance against *F. graminearum* and *F. culmorum* in wheat and barley [3]. Resistance mechanisms have been elucidated by using wheat/barley genotypes with various levels of resistance, classified as susceptible, intermediate and resistant. However, in most of the studies, unrelated germoplasms are compared, leading to a confusing interpretation of the data delivered, since the differences in the metabolic profiles may actually result from the cultivar background [3]. Thus, the use of near isogenic lines (NILs) that differ in QTL conditioning FHB, is suggested to be the best approach to simplify the complexity, and allow to reach conclusive evidence related to resistance functions [18].

As for the comparison, mock-inoculated versus pathogen-inoculated plants is considered the best approach to highlight differences. Gunnaiah et al. [19] instead designed a different experiment in order to elucidate the host biochemical resistance to FHB spread in response to trichothecene producing and nonproducing isolates of *F. graminearum*. The two *F. graminearum* strains differed in the loss of function of Tri5 gene [19]. In addition to *F. graminearum* inoculation, Warth et al. [15] also used DON injection into the middle florets of spikelets to decipher the mechanism of plant resistance to the toxin. Experiments have been performed in field conditions [27], under greenhouse [16–18, 33, 36] with computer-controlled settings for light, temperature and relative air humidity [15] and more recently, in environmental controlled growth chamber [34]. All these approaches are summarized in **Table 1** together with the extraction and detection methodologies applied, the plants used and the main classes of metabolites identified by the authors so far.

5. Elucidating FHB resistance mechanisms by metabolomics

Plant resistance to *Fusarium* Head Blight and related mycotoxin accumulation has been described through five major types of mechanism, mainly described for wheat and further applied to other cereals. These mechanisms are often host-specific, thus requiring plant-specific elucidation studies. Type I resistance is related to initial infection of the floret in wheat and barley, and of the silk in maize [37]. The spreading of infection is then limited by type II and type III resistance. Type IV resistance is related with tolerance and ability to maintain yields, and type V resistance gathers all mechanisms of resistance to mycotoxin accumulation [38–40]. According to Boutigny et al. [41], type V-1 represents resistance to toxin accumulation operated by metabolic biotransformation [42, 43], while type V-2 corresponds to resistance due to the inhibitory effect of mycotoxin biosynthesis exerted by plant endogenous compounds. Metabolomics has been exploited so far in this field for the comparison of metabolite composition of resistant and susceptible varieties upon *Fusarium* infection, allowing for the definition of a large set of compounds potentially involved in FHB modulation [3, 15–19]. Among those, fatty acids and compounds thereof have been found to be involved in the plant-pathogen signalling system, while terpenoids and phenylpropanoids take part to cell wall reinforcement, show antifungal properties and may interfere with mycotoxin biosynthesis [3]. Generally, the workflow of markers identification comprises the following steps: (1) marker identification based on accurate mass (MS), isotopic pattern and MS/MS pathway, (2) off- or online database searching and (3) data interpretation. These markers can be tentatively identified without analytical standards, or unambiguously identified using analytical standards. The identification of markers usually represents the last step within metabolomics studies. This is crucial in order to understand the metabolite pathway, since they can be interesting intermediates or final secondary metabolites. In this particular topic of mycotoxin contamination, hundreds of metabolites related to FHB resistance have been putatively identified so far by metabolomics strategies [4]. It was already mentioned that the number and chemical structures of metabolites significantly vary according to the experimental design and the applied analytical strategy.

Biomarker metabolites of resistance can be further subclassified according to their function. Those metabolites, whose abundance was increased in both resistant and susceptible cultivars, following pathogen inoculation as compared with those inoculated with water, were referred as pathogenesis-related (PR) metabolites [44]. Accordingly, metabolites that were significantly higher in resistant cultivars than in susceptible one were designated as resistance-related (RR) metabolites.

Among RR metabolites, some of them have been demonstrated as constitutive, while others are induced upon fungal infection [16, 17]. Among them, resistance-indicator metabolites [3, 16, 17] include modified mycotoxins such as DON, DON-3Glc and the other DON-biotransformation products (**Figure 3**). Following wheat inoculation by *Fusarium*, DON is spread within spike, and the host counteracts mycotoxins by conjugating them to endogenous metabolites (i.e., by glycosylation, acylation, conjugation to amino acids and glutathione). Thus, all the modified forms are designed as resistance indicators, since they indicate that the plant is reacting against the infection also by converting mycotoxins into their less toxic forms. According to the literature [19, 32], the chemical defense against fungal pathogens including DON producing *Fusarium* species is linked to three main mechanisms of resistance: cell wall reinforcement through the deposition of lignin, production of antimicrobial compounds and specific induction of defense signalling pathways. As reported by Gunnaiah et al. [18] among the metabolites reported as involved in plant response to FHB in soft wheat, the main chemical groups are phenylpropanoids, and terpenoids, followed by amino acid derivatives. On the other hand, when functional properties are considered, the majority of resistance-related metabolites showed an antimicrobial activity, followed by cell wall strengthening properties.

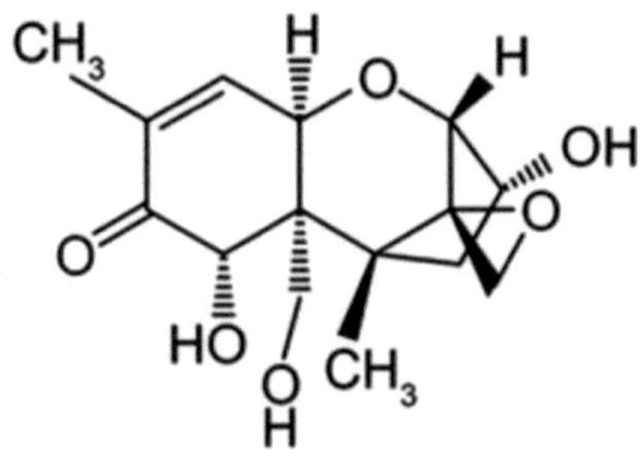

Figure 3. Chemical structure of deoxynivalenol (DON).

Phenylpropanoids such as flavonoids and phenolic acids have been frequently described for their contribution to plant defense mechanisms. Their activity is exerted either through direct interference with the fungus, or through the reinforcement of plant structural components acting as a mechanical barrier [45, 46]. Flavonoids, especially flavones, flavonones and isoflavonoids, lignans and other phenolic compounds were induced in Sumai-3 as antimicrobial agents, following *F. graminearum* inoculation. This is mainly due to their antioxidant activity leading to the neutralization of ROS, produced under biotic stress. A similar profile was identified upon *F. graminearum* inoculation in barley cultivars [16, 47] and in wheat [18]. In addition, phenolic acids have been reported as inhibitory agents towards mycotoxin biosynthesis in vitro [48, 49]. Among phenolic acids, hydroxycinnamic acid (HCA) derivatives, such as ferulic and caffeic acids, have been reported as important contributors to FHB resistance [4], probably on account of the high antioxidant properties [50].

Among HCAs, chlorogenic acid has been reported as a potential resistance factor in different pathosystems [49, 51, 52]. Concerning the cell wall reinforcement, hydroxycinnamic acid

http://dx.doi.org/10.5772/65647

amides (HCAAs) are deposited as cell wall appositions at the inner side of plant cell walls after cross-linking with polysaccharides, lignin and suberin [27]. These HCAAs are synthesized by condensation of hydroxycinnamoyl-CoA thioesters with aromatic amines (e.g., spermidine, spermine, tyramine) originated from aromatic amino acids. Thus, the involvement of amino acids in resistance to *Fusarium* may also be related to their role as a precursor of cell wall-bound HCAAs. Among those identified so far, N-caffeoylputrescine, 4-coumaroyl-3-hydroxyagmatine and feruloyl-serotonin are significantly upregulated upon *F. graminearum* infection in the resistant cultivar Sumai-3 [27]. With regards to the differences in terpenoid profile, Sumai-3 was characterized by a higher amount of syringyl lignin precursors like sinapoyl alcohol and sinapaldehyde, and glucose conjugate of sinapoyl alcohol, syringing [27]. Lignin results from monolignol glucosides' polymerizations and lead to a reinforced cell wall that is more resistant to fungal cell wall degradation enzymes [4].

Moreover, changes in the cell wall polysaccharides following infection were described by Cuperlovic-Culf et al. [24]. Large increase in concentration of sugars and inositols was found in all wheat varieties, particularly for Sumai-3, indicating an attempt at creation of cell wall barrier for *F. graminearum* penetration. In addition, fatty acids were also suggested to participate in resistance as physical barrier to pathogen ingress through their role in cuticle formation [4]. As far as the involvement of resistance related metabolites—mainly lipids—in the plant signalling pathways, significant results are summarized in the last part of this review.

6. The role of lipids in the plant-pathogen cross-talk

Increasing evidence indicates that lipid signalling is an integral part of the complex regulatory network in plant response to stress/infection. Modifications of membrane lipids produce different classes of signalling messengers, such as phosphatidic acid (PA), diacylglycerol (DAG), DAG pyrophosphate (DAGPP), lysophospholipids, free fatty acids (FFAs), oxylipins, phosphoinositides and inositol polyphosphates. Lipidomic approaches were developed to investigate in depth the plant-pathogen cross-talk, demonstrating a close relationship between the modification of the pathogen oxylipin profile with the mycotoxin synthesis [28].

Among metabolites associated with fatty acid metabolic pathways, a number of compounds have been identified for their potential contribution to cereal resistance towards FHB [53]. Fatty acids and their derivatives play significant role in plant defense against pathogens. Among their functions, they contribute to basal immunity, gene-mediated and systemic acquired resistance in plants. In addition, fatty acids are involved in the plant defense signalling pathway, through the formation of important mediators such as oxylipins and jasmonates. The unsaturated C18:1, C18:2 and C18:3 fatty acids, namely oleic, linoleic and linolenic acid, are often described as involved into defense mechanisms against fungal pathogens [47, 54, 55] and able to modulate mycotoxin production [55, 56]. The antimicrobial activity is probably due to their role in modulating ROS production, and in cuticle formation, which constitutes a physical barrier to pathogen infection [57]. In addition, they are precursors of the plant oxylipin pathway, which moves from the enzymatic formation of hydroperoxides, carried out by lipoxygenase (LOX) [58]. Distinct LOX isoforms, referred as 9-LOX and 13-LOX, preferentially add a hydroxyl moiety at C9 or C13 position of the fatty acid backbone, leading therefore to 9- and 13-hydroperoxides, respectively. These compounds act then as substrates for the two distinct biosynthetic cascades, with the formation of approximately 150 known oxylipins including hydroxy-, oxo- or keto-fatty acids, green leaf volatiles (GLVs) and jasmonic acid (JA) [59]. Jasmonates originate from 13-LOX products, while 9-LOX products lead to less-known metabolites known as defense factors in response to fungal attack [60]. Jasmonic acid and methyl jasmonate are well known for their roles as plant stress hormones. They cause programmed cell death activation, the production of ROS and the deposit of wax layers on plant tissues [61]. Jasmonates play, in addition, an active role in the regulation of the phenylpropanoids pathway [62], exhibit antimicrobial properties towards toxigenic fungi [47, 60] and modulate mycotoxin accumulation [63, 64].

Besides these functions, jasmonates were proved to activate glucosyltransferase in *Arabidopsis thaliana* and barley [65]. This is a key enzyme activity involved in a DON detoxification pathway that transforms DON into less phytotoxic DON-3-Glc. Several metabolomic studies have highlighted the involvement of jasmonic acid [15–19, 33] in resistance to DON-producing *Fusarium* species. While the physiological function of jasmonates has been well described over the last years, little is known about other 9-LOX-derived compounds. Recent studies demonstrated that 9-oxylipins contribute to maize susceptibility or resistance to fungal pathogens, in a pathosystem-dependent way [61]. Several studies, indeed, suggested that mycotoxin accumulation is modulated by host oxylipins. In particular, linoleic acid and 9-oxylipins seem to be conserved signal molecules modulating mycotoxin biosynthesis, fungal sporulation and other aspects of fungal differentiation processes [54]. The effects of mutation of LOX gene were often studied in maize, observing that inactivation of the 9-LOX gene led to an increased susceptibility of maize to *Aspergillus flavus*, *A. nidulans* and *F. verticillioides* [66–68]. Similarly, modification of LOX genes led to a modulation of fumonisin production in the maize—*F. verticillioides* pathosystem [69, 70]. The deep involvement of oxylipins in the intense cross-talk between host and pathogen has still to be clarified. Endogenous fungal oxylipins are known indeed for supporting host colonization, as well as mycotoxin biosynthesis. Some authors suggest the possible interaction between fungal oxylipins and plant GPCRs, transmembrane-proteins or receptor-like kinases, for host manipulation.

7. Conclusions

A metabolomics approach may support the quick growth of this relatively new field of research, allowing for a better understanding of the changes occurring in the plant and pathogen metabolites upon interaction. In principle, analytical methods developed have demonstrated significant advances in sensitivity, robustness, flexibility and discrimination power in order to build successfully statistical models, and subsequent marker identification. Increasing evidence indicates that lipid signalling is an integral part of the complex regulatory network in plant response to stress/infection. Modifications of membrane lipids produce different classes of signalling messengers, such as phosphatidic acid, diacylglycerol pyrophosphate, lysophospholipids, free fatty acids, oxylipins, phosphoinositides and inositol polyphosphates. Lipidomic approaches can be developed to investigate in depth the plant-pathogen cross-talk, demonstrating a close relationship between the modification of the pathogen oxylipin profile with the mycotoxin synthesis. Therefore, metabolomics approaches will provide new solutions to old problems. In fact, the early detection of mycotoxins and smart detoxifications can be performed by metabolomics strategies for the first time, and these approaches can fill the gap in order to answer these questions and go a step further.

Acknowledgements

Josep Rubert thanks the Generalitat Valenciana (Conselleria d'Educació, Cultura i Esport) for the VALi+d postdoctoral fellowship 'Contractació de personal investigador en formació en fase postdoctoral 2014' [grant number APOSTD/2014/120].

Author details

Laura Righetti[1], Chiara Dall'Asta[1], Jana Hajslova[2] and Josep Rubert[2*]

*Address all correspondence to: rubertbj@vscht.cz

1 Department of Food Science, University of Parma, Parma, Italy

2 Department of Food Analysis and Nutrition, Faculty of Food and Biochemical Technology, University of Chemistry and Technology, Prague, Czech Republic

References

[1] Bennett JW, Klich M. Mycotoxins. Clin. Microbiol. Rev.. 2003; 16(3): 497–516. DOI: 10.1128/CMR.16.3.497-516.2003.

[2] Hussein HS, Brasel JM. Toxicity, metabolism, and impact of mycotoxins on humans and animals. Toxicology. 2001; 167(2): 101–134. DOI: 10.1016/S0300-483X(01)00471-1.

[3] Bai G, Shaner G. Management and resistance in wheat and barley to Fusarium head blight. Annu. Rev. Phytopathol. 2004; 42: 135–161. DOI: 10.1146/annurev.phyto.42.040803.

[4] Gauthier L., Atanasova-Penichon V., Chéreau S., Richard-Forget F. Metabolomics to decipher the chemical defense of cereals against *Fusarium graminearum* and deoxynivalenol accumulation. Int. J. Mol. Sci. 2015; 16: 24839–24872. DOI: 10.3390/ijms161024839.

[5] International Agency for Research on Cancer (IARC). Aflatoxins. In Some Traditional Herbal Medicines, Some Mycotoxins, Naphthalene and Styrene. IARC Monogr. Eval. Carcinog. Risks Hum. 2002; 82: 1–556.

[6] International Agency for Research on Cancer (IARC). Ochratoxin A. In Some Naturally Occurring Substances: Food Items and Constituents, Heterocyclic Aromatic Amines and Mycotoxins. IARC Monogr. Eval. Carcinog. Risks Hum. 1993; 56: 489–521.

[7] Lemmens M, Scholz U, Berthiller F, Dall'Asta C, Koutnik A, Schuhmacher R, Adam G, Buerstmayr H, Mesterházy A, Krska R, Ruckenbauer P. The ability to detoxify the mycotoxin deoxynivalenol colocalizes with a major quantitative trait locus for Fusarium head blight resistance in wheat. Mol. Plant Microbe Interact.. 2005; 18(12): 1318–1324. DOI: 10.1094/MPMI-18-1318.

[8] Logrieco A, Bottalico A, Mulé G, Moretti A, Perrone G. Epidemiology of toxigenic fungi and their associated mycotoxins for some Mediterranean crops. Eur. J. Plant Pathol. 2003; 109(7): 645–667. DOI: 10.1023/A:1026033021542.

[9] Turner NW, Bramhmbhatt H., Szabo-Vezse M, Poma A, Coker R, Piletsky SA. Analytical methods for determination of mycotoxins: an update (2009–2014). Anal. Chim. Acta 2015; 901: 12–33. DOI: 10.1016/j.aca.2015.10.013.

[10] Rubert J, Dzuman Z, Vaclavikova M, Zachariasova M, Soler C, Hajslova J. Analysis of mycotoxins in barley using ultra high liquid chromatography high resolution mass spectrometry: comparison of efficiency and efficacy of different extraction procedures. Talanta 2012; 99: 712–719. DOI: 10.1016/j.talanta.2012.07.010.

[11] Berthiller F, Brera C, Crews C, Iha M-H, Krska R, Lattanzio V-M-T, MacDonald S, Malone R-J, Maragos C, Solfrizzo M, Stroka J, Whitaker TB. Developments in mycotoxin analysis: an update for 2014–2015. WMJ. 2016; 9(1): 5–29. DOI: http://dx.doi.org/10.3920/WMJ2015.1998.

[12] Dzuman Z, Zachariasova M, Veprikova Z, Godula M, Hajslova J. Multi-analyte high performance liquid chromatography coupled to high resolution tandem mass spectrometry method for control of pesticide residues, mycotoxins, and pyrrolizidine alkaloids. Anal. Chim. Acta 2015; 863: 29–40. DOI: 10.1016/j.aca.2015.01.021.

[13] Zachariasova M, Cuhra P, Hajslova J. Cross-reactivity of rapid immunochemical methods for mycotoxins detection towards metabolites and masked mycotoxins: the current state of knowledge. WMJ. 2014; 7: 449–464. DOI: http://dx.doi.org/10.3920/WMJ2014.1701.

[14] Fox G, Manley M. Applications of single kernel conventional and hyperspectral imaging near infrared spectroscopy in cereals. J. Sci. Food Agric. 2014; 94(2): 174–179. DOI: 10.1002/jsfa.6367.

[15] Warth B, Parich A, Bueschl C, Schoefbeck D, Neumann NKN, Kluger B, Schuster K, Krska R, Adam G, Lemmens M, Schuhmacher R. GC-MS based targeted metabolic

profiling identifies changes in the wheat metabolome following deoxynivalenol treatment. Metabolomics. 2015; 11: 722–738. DOI: 10.1007/s11306-014-0731-1.

[16] Bollina V, Kumarasswamy GK, Kushalappa AC, Choo TM, Dion Y, Rioux S, Faubert D, Hamzehzarghani H. Mass spectrometry-based metabolomics application to identify quantitative resistance-related metabolites in barley against Fusarium head blight. Mol. Plant Pathol. 2010; 11(6): 769–782. DOI: 10.1111/J.1364-3703.2010.00643.X.

[17] Bollina V, Kushalappa AC, Choo TM, Dion Y, Rioux S. Identification of metabolites related to mechanism of resistance in barley against Fusarium gramineuarum, based on mass spectrometry. Plant Mol. Biol. 2011; 77: 355–370. DOI: 10.1007/s11103-011-9815-8.

[18] Gunnaiah R, Kushalappa AC, Duggavathi R, Fox S, Somers D. Integrated metabolo-proteomic approach to decipher the mechanisms by wich wheat QTL (Fhb1) contributes to resistance against Fusarium graminearum. PLoS One 2012; 7(7): e40695. DOI: 10.1371/journal.pone.0040695.

[19] Gunnaiah R, Kushalappa AC. Metabolomics deciphers the host resistance mechanisms in wheat cultivar Sumai-3, against trichotecene producing and non-producing isolates of Fusarium graminearum. Plant Physiol. Biochem. 2014; 83: 40–50. DOI: 10.1016/j.plaphy.2014.07.002.

[20] Geddes J, Eudes F, Laroche A, Selinger LB. Differential expression of proteins in response to the interaction between the pathogen Fusarium graminearum and its host, Hordeum vulgare. Proteomics. 2008; 8(3): 545–554. DOI: 10.1002/pmic.200700115.

[21] Kluger B, Bueschl C, Lemmens M, Michlmayr H, Malachova A, Koutnik A, Maloku I, Berthiller F, Adam G, Krska R, Schuhmacher R. Biotransformation of the mycotoxin deoxynivalenol in Fusarium resistant and susceptible near isogenic wheat lines. PLoS One. 2015; 10(3): e0119656. DOI: 10.1371/journal.pone.0119656.

[22] Berthiller F, Werner U, Sulyok M, Krska R, Hauser MT, Schuhmacher R. Liquid chromatography coupled to tandem mass spectrometry (LC-MS/MS) determination of phase II metabolites of the mycotoxin zearalenone in the model plant Arabidopsis thaliana. Food Addit. Contam. 2006; 23(11): 1194–1200. DOI: 10.1080/02652030600778728.

[23] Fiehn O. Metabolomics—the link between genotypes and phenotypes. Plant Mol. Biol. 2002; 48(1–2): 155–171. DOI: 10.1023/A:1013713905833.

[24] Rubert J, Zachariasova M, Hajslova J. Advances in high-resolution mass spectrometry based on metabolomics studies for food – a review. Food Addit. Contam. Part A Chem. Anal. Control Expo. Risk Assess. 2015; 32(10): 1685–1708. DOI: 10.1080/19440049.2015.1084539.

[25] Del Fiore A, Reverberi M, Ricelli A, Pinzari F, Serranti S, Fabbri AA, Bonifazi G, Fanelli C. Early detection of toxigenic fungi on maize by hyperspectral imaging analysis. Int. J. Food Microbiol. 2010; 144(1): 64–71. DOI: 10.1016/j.ijfoodmicro.2010.08.001.

[26] Berardo N, Pisacane V, Battilani P, Scandolara A, Pietri A, Marocco A. Rapid detection of kernel rots and mycotoxins in maize by near-infrared reflectance spectroscopy. J. Agric. Food Chem.. 2005; 53(21): 8128–8134. DOI: 10.1021/jf0512297.

[27] Cajka T, Vaclavikova M, Dzuman Z, Vaclavik L, Ovesna J, Hajslova J. Rapid LC–MS-based metabolomics method to study the Fusarium infection of barley. J. Sep. Sci. 2014; 37; 912–919. DOI: 10.1002/jssc.201301292.

[28] Scala V, Camera E, Ludovici M, Dall'Asta C, Cirlini M, Giorni P, Battilani P, Bello C, Fabbri AA, Fanelli C, Reverberi M. Fusarium verticillioides and maize interaction in vitro: relationship between oxylipin cross-talk and fumonisin synthesis. World Mycotoxin J. 2013; 6(3): 343–351. DOI: http://dx.doi.org/10.3920/WMJ2012.1527.

[29] Ludovici M, Ialongo C, Reverberi M, Beccaccioli M, Scarpari M, Scala M. Quantitative profiling of oxylipins through comprehensive LC-MS/MS analysis of Fusarium verticillioides and maize kernels. Food Addit. Contam. Part A. 2014; 31(12): 2026–2033. DOI: 10.1080/19440049.2014.968810.

[30] Giorni P, Dall'Asta C, Reverberi M, Scala V, Ludovici M, Cirlini M, Galaverna G, Fanelli C, Battilani P. Open field study of some *Zea mays* hybrids, lipid compounds and fumonisins accumulation. Toxins. 2015; 7: 3657–3670. DOI: 10.3390/toxins7093657.

[31] Nathanail AV, Varga E, Meng-Reiterer J, Bueschl C, Michlmayr H, Malachova A, Fruhmann P, Jestoi M, Peltonen K, Adam G, Lemmens M, Schuhmacher R, Berthiller F. Metabolism of the Fusarium mycotoxins T-2 toxin and HT-2 toxin in wheat. J. Agric. Food Chem. 2015; 63: 7862–7872. DOI: 10.1021/acs.jafc.5b02697.

[32] Kushalappa AC, Gunnaiah R. Metabolo-proteomics to discover plant biotic stress resistance genes. Trends Plant Sci. 2013; 11(9): 522–531. DOI: 10.1016/j.tplants. 2013.05.002.

[33] Browne RA, Brindle KM. H NMR-based metabolite profiling as potential selection tool for breeding passive resistance against Fusarium head blight (FHB) in wheat. Mol. Plant Pathol. 2007; 8(4): 401–410. DOI: 10.1111/j.1364-3703.2007.00400.x.

[34] Cuperlovic-Culf M, Wang L, Forseille L, Boyle K, Merkley N, Burton I, Fobert PR. Metabolic marker panels to response to Fusarium Head Blight infection in different wheat varieties. PLoS One. 2016; 11(4): e0153642. DOI: 10.1371/journal.pone.0153642.

[35] Paranidharan V, Abu-Nada Y, Hamzehzarghani H, Kushalappa AC, Mamer O, Dion Y, Rioux S, Comeau A, Choiniere L. Resistance-related metabolites in wheat against *Fusarium graminearum* and the virulence factor deoxynivalenol (DON). Botany. 2008; 86: 1168–1179. DOI: 10.1139/B08-052.

[36] Nussbaumer T, Warth B, Sharma S, Ametz C, Bueschl C, Parich A, Pfeifer M, Siegwart G, Steiner B, Lemmens M, Schuhmacher R, Buerstmayr H, Mayer KFX, Kugler KG,

Schweiger W. Joint transcriptomic and metabolomic analyses reveal changes in the primary metabolism and imbalances in the subgenome orchestration in the bread wheat molecular response to Fusarium graminearum. G3-Genes Genom Genet. 2005; 5(12): 2579–2592. DOI: 10.1534/g3.115.021550.

[37] Schroeder HW, Christensen JJ. Factors affecting resistance of wheat to scab caused by Gibberella zeae. Phytopathology 1963; 53: 831–838.

[38] Miller JD, Young JC, Sampson DR. Deoxynivalenol and Fusarium head blight resistance in spring cereals. J. Phytopathol. 1985; 113: 359–367.

[39] Mesterhazy A. Types and components of resistance to Fusarium head blight of wheat. Plant Breeding. 1995; 114: 337–386.

[40] Mesterhazy A. Role of deoxynivalenol in aggressiveness of *Fusarium graminearum* and *F. culmorum* and in resistance to Fusarium head blight. Eur. J. Plant Pathol.. 2002; 108: 675–684. DOI: 10.1023/A:1020631114063.

[41] Boutigny AL, Richard-Forget F, Barreau C. Natural mechanisms for cereal resistance to the accumulation of Fusarium trichothecenes. Eur. J. Plant Pathol. 2008; 121: 411–423. 10.1007/s10658-007-9266-x.

[42] Karlovsky P. Biological detoxification of the mycotoxin deoxynivalenol and its use in genetically engineered crops and feed additives. Appl. Microbiol. Biotechnol. 2011: 91; 491–504. 10.1007/s00253-011-3401-5.

[43] De Boevre M, Landschoot S, Audenaert K, Maene P, Mavungu DJD, Eeckhout MD. Occurrence and within field variability of Fusarium mycotoxins and their masked forms in maize crops in Belgium. World Mycotoxin J. 2014; 7: 91–102. 10.3920/WMJ2013.1608.

[44] Hamzehzarghani H, Kushalappa AC, Dion Y, Rioux S, Comeau A, Yaylayan V, Marshall WD, Mather DE. Metabolic profiling and factor analysis to discriminate quantitative resistance in wheat cultivars against Fusarium head blight. Physiol. Mol. Plant Pathol. 2005; 66: 119–133. DOI: 10.1016/j.pmpp.2005.05.005.

[45] Lattanzio V, Lattanzio VMT, Cardinali A. Role of phenolics in the resistance mechanisms of plants against fungal pathogens and insects. Phytochemistry 2006; 23–67.

[46] Siranidou E, Kang Z, Buchenauer H. Studies on symptom development, phenolic compounds and morphological defence responses in wheat cultivars differing in resistance to Fusarium Head Blight. J. Phytopathol. 2002; 150: 200–208. DOI: 10.1046/j.1439-0434.2002.00738.x.

[47] Kumaraswamy GK, Bollina V, Kushalappa AC, Choo TM, Dion Y, Rioux S, Mamer O, Faubert, D. Metabolomics technology to phenotype resistance in barley against gibberella zeae. Eur. J. Plant Pathol. 2011; 130: 29–43. DOI: 10.1007/s10658-010-9729-3.

[48] Boutigny AL, Barreau C, Atanasova-Penichon V, Verdal-Bonnin MN, Pinson-Gadais L, Richard-Forget F. Ferulic acid, an efficient inhibitor of type B trichothecene biosynthesis

and Tri gene expression in Fusarium liquid cultures. Mycol. Res.. 2009; 113(6–7): 746–753. DOI: 10.1016/j.mycres.2009.02.010.

[49] Atanasova-Penichon V, Bernillon S, Marchegay G, Lornac A, Pinson-Gadais L, Ponts N, Zehraoui E, Barreau C, Richard-Forget F. Bioguided isolation, characterization, and biotransformation by Fusarium verticillioides of maize kernel compounds that inhibit fumonisin production. Mol. Plant-Microbe Interact. 2014; 27: 1148–1158. DOI: 10.1094/MPMI-04-14-0100-R.

[50] Ponts N, Pinson-Gadais L, Boutigny AL, Barreau C, Richard-Forget F. Cinnamic derived acids significantly affect *Fusarium graminearum* growth and in vitro synthesis of type B trichothecenes. Phytopathology 2011; 101: 929–934. DOI: 10.1094/PHYTO-09-10-0230.

[51] Villarino M, Sandin-Espana P, Melgarejo P, De Cal A. High chlorogenic and neochlorogenic acid levels in immature peaches reduce *Monilinia laxa* infection by interfering with fungal melanin biosynthesis. J. Agric. Food Chem. 2011; 59: 3205–3213. DOI: 10.1021/jf104251z.

[52] Wojciechowska E, Weinert CH, Egert B, Trierweiler B, Schmidt-Heydt M, Horneburg B, Graeff-Honninger S, Kulling SE, Geisen R. Chlorogenic acid, a metabolite identified by untargeted metabolome analysis in resistant tomatoes, inhibits the colonization by *Alternaria alternata* by inhibiting alternariol biosynthesis. Eur. J. Plant Pathol. 2014; 139: 735–747. DOI: 10.1007/s10658-014-0428-3.

[53] Kachroo A, Kachroo P. Fatty acid-derived signals in plant defense. Annu. Rev. Phytopathol. 2009; 47: 153–176. DOI: 10.1146/annurev-phyto-080508-081820.

[54] Burow GB, Nesbitt TC, Dunlap J, Keller NP. Seed lipoxygenase products modulate Aspergillus mycotoxin biosynthesis. Mol. Plant Microbe Interact. 1997; 10: 380–387. doi.org/10.1094/MPMI.1997.10.3.380.

[55] Tiwari RP, Mittal V, Singh G, Bhalla TC, Saini SS, Vadehra DV. Effect of fatty-acids on aflatoxin production by *Aspergillus parasiticus*. Folia Microbiol. 1986; 31: 120–123. DOI: 10.1007/bf02926829.

[56] Walters D, Raynor L, Mitchell A, Walker R, Walker K. Antifungal activities of four fatty acids against plant pathogenic fungi. Mycopathologia. 2004; 157: 87–90. DOI: 10.1023/B:MYCO.0000012222.68156.2c.

[57] Yaeno T, Matsuda O, Iba K. Role of chloroplast trienoic fatty acids in plant disease defense responses. Plant J. 2004; 40: 931–941. DOI: 10.1111/j.1365-313X.2004.02260.x.

[58] Feussner I, Wasternack C. The lipoxygenase pathway. Annu. Rev. Plant Biol. 2002; 53: 275–297. DOI: 10.1146/annurev.arplant.53.100301.135248.

[59] Mosblech A, Feussner I, Heilmann I. Oxylipins: structurally diverse metabolites from fatty acid oxidation. Plant Physiol. Biochem. 2009; 47: 511–517. DOI: 10.1016/j.plaphy.2008.12.011.

[60] Goodrich-Tanrikulu M, Mahoney NE, Rodriguez SB. The plant growth regulator methyl jasmonate inhibits aflatoxin production by *Aspergillus flavus*. Microbiology 1995; 141: 2831–2837. DOI: 10.1099/13500872-141-11-2831.

[61] Gao XQ, Kolomiets MV. Host-derived lipids and oxylipins are crucial signals in modulating mycotoxin production by fungi. Toxin Rev. 2009; 28: 79–88. DOI: 10.1080/15569540802420584.

[62] Kazan K, Manners JM. Jasmonate signaling: toward an integrated view. Plant Physiol. 2008; 146: 1459–1468. DOI: 10.1104/pp.107.115717.

[63] Meimaroglou DM, Galanopoulou D, Flouri F, Panagiota M. The plant growth regulator methyl jasmonate inhibits aflatoxin B1 production by *Aspergillus parasiticus* in caper. Int. J. Nutr. Food Sci. 2015: 141: 10–17. DOI: 10.1099/13500872-141-11-2831.

[64] Vergopoulou S, Galanopoulou D, Markaki P. Methyl jasmonate stimulates aflatoxin B1 biosynthesis by *Aspergillus parasiticus*. J. Agric. Food Chem. 2001; 49: 3494–3498. DOI: 10.1021/jf010074+.

[65] Poppenberger B, Berthiller F, Lucyshyn D, Sieberer T, Schuhmacher R, Krska R, Kuchler K, Glossl J, Luschnig C, Adam G. Detoxification of the Fusarium mycotoxin deoxynivalenol by a UDP-glucosyltransferase from *Arabidopsis thaliana*. J. Biol. Chem. 2003; 27: 47905–47914. DOI: 10.1074/jbc.M307552200.

[66] Gao X, Brodhagen M, Isakeit T, Brown SH, Göbel C, Betran J, Feussner I, Keller NP, Kolomiets MV. Inactivation of the lipoxygenase ZmLOX3 increases susceptibility of maize to *Aspergillus* spp. Mol. Plant Microbe Interact. 2009; 22: 222–231. DOI: 10.1094/MPMI-22-2-0222.

[67] Fabbri AA, Fanelli C, Panfili G, Passi S, Fasella P. Lipoperoxidation and aflatoxin biosynthesis by *Aspergillus parasiticus* and *A. flavus*. J. Gen. Microbiol. 1983; 129: 3447–3452.

[68] Fanelli C, Fabbri AA. Relationship between lipids and aflatoxin biosynthesis. Mycopathologia. 1989; 107: 115–120. DOI: 10.1007/BF00707547.

[69] Brodhagen M, Keller NP. Signalling pathways connecting mycotoxin production and sporulation. Mol. Plant Pathol. 2006; 7: 285–301. DOI: 10.1111/j.1364-3703.2006.00338.x.

[70] Nobili C, D'Angeli S, Altamura MM, Scala V, Fabbri AA, Reverberi M, Fanelli C. ROS and 9-oxylipins are correlated with deoxynivalenol accumulation in the germinating caryopses of *Triticum aestivum* after *Fusarium graminearum* infection. Eur. J. Plant Pathol. 2014; 139: 429–444. DOI: 10.1007/s10658-014-0401-1.

Chapter 7

Impact of Metabolomics in Symbiosis Research

Alba Chavez-Dozal and Michele K. Nishiguchi

Additional information is available at the end of the chapter

http://dx.doi.org/10.5772/66631

Abstract

In symbiotic associations, there is a constant molecular complexity that allows establishment and maintenance of the relationship. Metabolomic profiles have enabled researchers to explain symbiotic associations in terms of their underlying molecules and interactions between the symbiotic partners. In this review, we have selected studies on symbioses as examples that have helped to explain the metabolic integration of bacterial symbionts and their hosts in an effort to understand the molecular fingerprint of animal-microbial symbioses.

Keywords: symbiosis, mutualism, metabolomics, co-clustering analysis

1. Introduction

The intimate association between two organisms is a very complex biological phenomenon; nevertheless, it is a very common way of life for every living organism on Earth. Symbiotic associations with one or many phylogenetically different organisms provide a fascinating view into how symbionts adapt and co-evolve. As Chaston and Douglas beautifully described in their comprehensive review [1], the omics revolution has transformed our ability to understand symbiotic associations at the molecular level. Researchers have adopted multiple techniques with great fervor in an effort to decipher the basis and complexity of symbiotic associations. Until recent years, the molecular pathways of symbiotic associations could only be studied in the context of genetic changes (transcriptomic studies) and protein profiles (proteomics); however, it is very likely that the establishment of a mutualistic association involved multiple evolutionary changes in the biochemistry and metabolic network of all the partners involved in the symbiosis [1]. Omics biology brings challenges and opportunities; one of the recent advances is the ability to construct a molecular metabolic catalog of an organism within a symbiotic association.

Metabolomics refers to the analytical approach used to study different cell products ("chemical fingerprints") that help to understand the physiological state of an organism [2].

In this section, we provide a comprehensive description of four experiments where the approach of metabolomics was selected in a particular type of animal-microbial symbiosis, in order to answer specific questions in symbiosis research.

2. Exemplars of metabolomic approaches in symbiosis research

2.1. Inferring metabolic interactions in arbuscular mycorrhizal symbiosis

Our exemplar of metabolomics studies of microbe-plant interactions is a set of observations by Schweiger et al. [3] that describe species-specific leaf metabolic responses to arbuscular mycorrhiza (AM) [4]. Arbuscular mycorrhiza is a unique symbiotic association between root arbuscular mycorrhizal fungi (AMF) and plants [4]. This is an ancient and widespread association where the fungus improves water uptake to the host plant, and in return the fungus receives plant carbohydrates. The fungus is restricted to the roots of the plant; however, the biochemical pathways and the involvement of exchanged substances are reflected on systemic root tissues affecting the chemical composition of plant tissues (defined as "phyto-metabolome") [4].

Comparative studies conducted on five different plant-AMF associations demonstrate that foliar metabolome is highly plant-species-specific, with low degrees of conservation across species. The experimental design was crucial to the success of this analysis, with the metabolome analysis performed on leaves of five plant species exposed to the worldwide distributed AMF *Rhizophagus irregularis*. Furthermore, the study took into account the implications of metabolite fluctuation at different leaf developmental stages and plant-reproductive status. Additionally, mycorrhizal plants were compared with control plants that received a sterilized inoculum. The results from this study indicate the high specificity of plant metabolome responses to the same AMF colonization; among the most striking findings indicate that metabolomics responses related to phosphate uptake, citric acid cycle, and amino acids were species-specific [3]. **Figure 1** summarizes the most important findings of this interesting study.

2.2. Metabolomic profile of the ryegrass-endophyte symbiosis

Along the lines of microbe-plant interactions, there is an interesting study conducted by Cao et al. [5] that is of particular relevance for symbiosis research. The metabolomics profile of perennial ryegrass (*Lolium perenne*) infected with endophytic fungus (*Neotyphodium lolii*) provided understanding of regulatory biochemical mechanisms for the production of beneficial alkaloids.

N. lolii is a naturally occurring fungus whose complete cycle occurs within perennial ryegrass. The fungus grows between the cells of the host plant drawing nutrients from it, and in return, the endophyte produces chemical compounds that provide resistance to drought, pests, and protection from overgrazing. Therefore, the aim of this study was to gather metabolomics information and combine it with microarray data in order to obtain a better understanding of

http://dx.doi.org/10.5772/66631

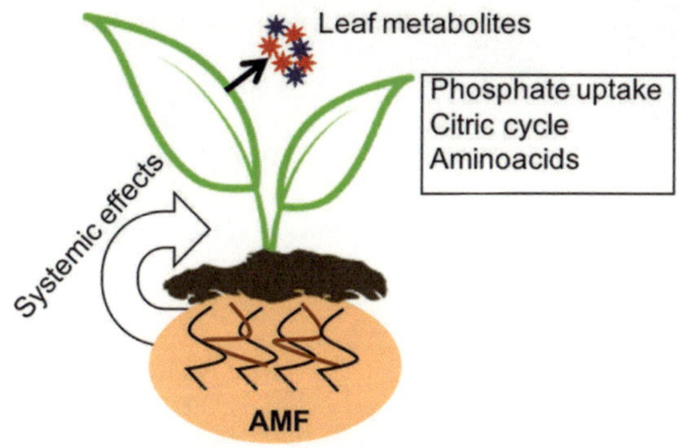

Figure 1. Summary of findings on the leaf phytometabolome when plants are exposed to the same arbuscular mycorrhizal fungi (AMF). Leaf metabolites detected included carbohydrates, organic acids, amino acids and derivatives, cyclic polyols, and sugar alcohols. Metabolites are differentially regulated primordially affecting the phosphate and citric acid cycles.

the biochemical mechanisms involved in the cross talk between partners, with the eventual purpose of achieving genetic manipulation of beneficial metabolite production (in particular manipulation of alkaloids).

Twenty-four perennial ryegrass samples comprising three tissue types (immature leaves, blades, and mature leaves) were examined of both endophyte-infected plants and endophyte free as a control. Targeted metabolomics analysis was used as the quantitative approach that provided identities of 70 metabolites based on the available databases of reference compounds. The use of targeted metabolomics in combination with microarray data provided better identification and classification Accuracy of compounds, as well as greater insights into the dynamics and fluxes of the newly identified metabolites. Results of this comprehensive study included the identification of accumulated alkaloids in the mature tissues of endophyte-infected ryegrass, and the co-clustering analysis of microarray data-identified genes with distinctive expression patterns which coincide with the pattern of alkaloid accumulation [5]. **Figure 2** summarizes the findings of this study. Results of this study indicate that co-clustering analysis is not a straightforward task no matter what kind of algorithm is used, and that the integration of transcriptomics and metabolomics can generate noisy data. However, this study demonstrated that co-cluster analysis could be a comprehensive choice to gain a more complete understanding of a complex biological system involving two entirely different taxa that are intertwined in their metabolic capabilities.

2.3. Metabolomic profile of symbiotic protection against pathogens

It is believed that specific strains from the gut microbiota can influence host immunity and protect from infection by pathogenic bacteria. One example is the early and prevalent gut colonizer *Bifidobacterium*, which is considered part of the healthy normal gut flora. It is believed that different strains of *Bifidobacterium* protect against enteropathogenic *Escherichia coli* O157:H7 infection in mice; however, the potential molecular and cellular mechanisms underpinning this protective effect are still under investigation [6].

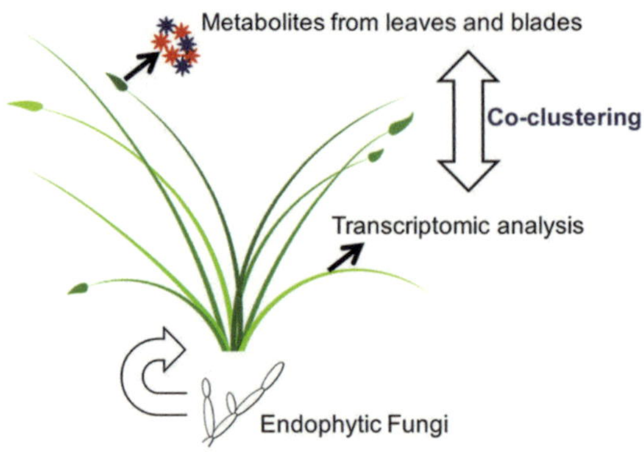

Figure 2. Summary of the study conducted by Cao et al. Co-clustering analysis of microarray and metabolomics data on endophytic-infected ryegrass indicate a set of genes and metabolites that are important for alkaloid production.

One study conducted by Fukuda et al. [7] used a combined "omics" strategy in an effort to gain a better understanding of the protective effect of *Bifidobacterium* over its mice host. Experiments designed comprised mice infected with different species of the symbiotic bacterium *Bifidobacterium* (including *B. longum* and *B. adolescentis*) and the pathogen *E. coli* O157:H7. The life span of co-infected mice was observed and transcriptomic and metabolomic profiles were conducted. **Figure 3** diagrams the experimental design of this study. This sophisticated analysis included a combination of sequencing, the platform used for metabolite detection was HPLC-MS (high-performance liquid chromatography-mass spectrometry) and for the analysis of products, the dataset was subjected to a multivariate analysis method named PLS (partial least squares) projection to latent structures. Typical data-processing flow included detection of signal peaks and normalization of dataset to generate a matrix of the products detected. For their statistical analysis, the method selected was PCA (principal component analysis) and CL (cluster analysis).

Results from this study indicate that mice bearing the strain *B. longum* survived, whereas those infected with *B. adolescentis* died. Metabolomic profiles between the two treatments revealed that the concentration of fatty acids (acetic acid in particular) was significantly elevated in those mice that survived *E. coli* infection. Furthermore, mice that survived showed an increased expression of genes involved in ATP-binding-cassette carbohydrate transporters [7]. Observations from the study suggest that the elevated production of acetic acid improved intestinal defense, thereby enhancing the barrier function of colon epithelial cells inhibiting the transport of *E. coli* toxins.

2.4. Metabolomics of a beneficial marine bacterium

The marine luminescent bacterium *Vibrio fischeri* establishes a symbiotic association with numerous sepiolid squids and monocentric fishes. *V. fischeri* infects a specialized light organ in the mantle (body) cavity of host squids and produces bioluminescence that is used by its host to avoid predation in a behavior known as counterillumination. In return, the squid host provides an enriched habitat for *Vibrio* to reproduce and to form bacterial communities of

http://dx.doi.org/10.5772/66631

monospecies biofilms. The ability of *V. fischeri* to form a biofilm in the light organ of its squid host plays a central role in establishment and maintenance of the symbiotic association. This interesting symbiotic association has been the center of attention of many researchers, and has been investigated for more than 25 years; however, as indicated for other examples of mutualistic associations, the molecular basis of the squid-*Vibrio* symbiosis is still obscure.

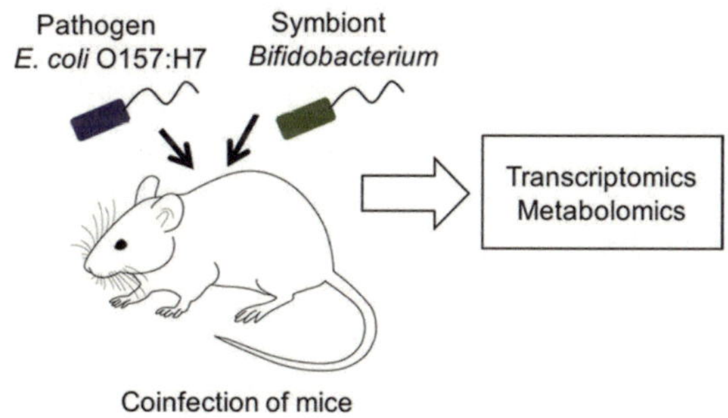

Figure 3. Summary of the experiment conducted by Fukoda et al. Mice were coinfected with beneficial strains of *Bifidobacterium* and the pathogenic strain of *Escherichia coli* O157:H7. Combined transcriptomic and metabolomic profiles revealed an increase in acetate and fructose transporters in those mice that survived lethal infection.

In a recent study conducted by Chavez-Dozal et al. [8], both proteomic and metabolomic profiles were performed in parallel in strains of *V. fischeri* in their biofilm form and compared to profiles of free-living (or planktonic) *V. fischeri* cells of the same strain. The main objective of this study was to obtain a comprehensive profile of the molecular components to provide the first meta-proteome profile of biofilms that are important for establishment of this mutualistic association. A summary of this study is illustrated in **Figure 4**.

Biofilms are a complex microbial community composed of cells encased within a self-produced exopolymeric matrix. Expression profiles of biofilm communities reveal the composition of the matrix, which include a combination of lipids, polysaccharides, proteins, and DNA [9, 10].

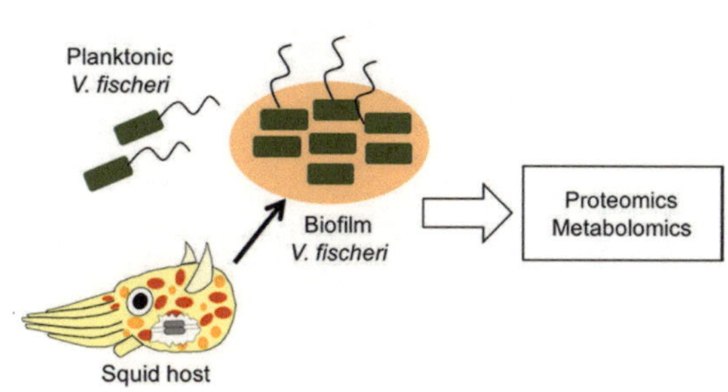

Figure 4. Summary of the experiment conducted by Chavez-Dozal et al. [5]. Proteomic and metabolomic profiles were performed in planktonic cells and biofilm communities of the same strain of *Vibrio fischeri*. Results revealed an upregulation of biofilm matrix components and molecules related to multiple stress responses.

Results of this study revealed a time-resolved picture of approximately 100 proteins and 200 metabolites present in the biofilm state of *V. fischeri*. The most important components found in this study include proteins, sugars, and molecules that form part of the exopolysaccharide matrix of biofilms; surprisingly, an increased concentration of intermediates of the glycolysis pathway was found to be prevalent during the biofilm state [8]. Results from this study suggest that molecules involved in the construction of the biofilm matrix are essential to bacterial community formation, a process that has been known to activate stress responses such as upregulation of alternative anaerobic pathways. The reported findings of this study have broad implications for *V. fischeri* ecology, since many of the symbiosis-regulated genes are not yet described. The combination of proteomics and metabolomics has therefore provided a link between protein regulation and function during different phases of the symbiosis, improving our understanding of the mechanisms that are important for successful host colonization.

3. Concluding remarks

Metabolomic approaches are increasingly selected for multiple purposes of symbiosis research. Although other "omic" approaches are needed to understand molecular function in symbiotic associations, the emerging use of metabolomics provides a new level of biochemical sophistication. The different examples provided in this mini review are only some of the pillar studies that included the use of either metabolomics or a combinational analysis of metabolomics with transcriptomics/proteomics of different mutualistic systems; however, many more studies are in progress using metabolomics profiles to define and characterize molecular and biochemical pathways that are important for establishment and persistence of symbiotic associations. The advancement of technologies that allows higher resolution of minute concentrations of proteins and their modulation will expand the area of metabolomics research and will enable a better perspective of the physiological state of organisms as single entities (otherwise known as the holobiome).

Author details

Alba Chavez-Dozal and Michele K. Nishiguchi*

*Address all correspondence to: nish@nmsu.edu

Department of Biology, New Mexico State University, Las Cruces, NM, USA

References

[1] Chaston J, Douglas AE. Making the most of "omics" for symbiosis research. Biol Bull; 2012;**223**:21–9. doi: 10.1111/j.1365-2133.2012.10859.x.

[2] Tyagi S, Raghvendra, Singh U, Kalra T, Munjal K. Applications of metabolomics, A systematic study of the unique chemical fingerprints: an overview. Int J Pharma Scie Rev and Res; 2010;**3**(1):019.

[3] Schweiger R, Baier MC, Persicke M, Muller C. High specificity in plant leaf metabolic responses to arbuscular mycorrhiza. Nat Comm; 20114;**5**:3886. doi: 10.1094/MPMI-05-14-0126-R.

[4] Schweiger R, Muller C. Leaf metabolome in arbuscular mycorrhizal symbiosis. Curr Opin Plant Biol; 2015;**26**:120–6. doi: 10.1016/j.pbi.2015.06.009.

[5] Cao M, Johnson L, Johnson R, Koulman A, Lane GA, Rasmussen S. Joint analyses of transcriptomic and Metabolomic data to probe ryegrass-endophyte symbiosis. In: Popay A, Thom E (eds). Proceedings of the 6th international symposium on fungal endophytes of grasses. Christchurch, New Zealand, December 2007.

[6] Sugahara H, Odamaki T, Fukuda S, Kato T, Xiao J, Abe F, Kikuchi J, Ohno H. Probiotic *Bifidobacterium longum* alters gut luminal metabolism through modification of the gut microbial community. Sci Rep; 2015;**5**:13548. doi:10.1038/srep13548.

[7] Fukuda S, Toh H, Hase K, Oshima K, Nakanishi Y, Yoshimura K, Tobe T, Clarke JM, Topping DL, Suzuki T, Taylor TD, Itoh K, Kikuchi J, Morita H, Hattori M, Ohno H. Bifidobacteria can protect from enteropathogenic infection through production of acetate. Nature; 2011;**469**:543–7. doi: 10.1038/nature09646.

[8] Chavez-Dozal A, Gorman C, Nishiguchi MK. Proteomic and metabolomics profiles demonstrate variation among free-living and symbiotic *Vibrio fischeri* biofilms. BMC Micro; 215;**15**:226. doi: 10.1186/s12866-015-0560-z.

[9] Hobley L, Harkins C, MacPhee CE, Stanley-Wall NR. Giving structure to the biofilm matrix: an overview of individual strategies and emerging common themes. FEMS Microbiol Rev; 2015;**39**(5):649–69. doi: 10.1093/femsre/fuv015.

[10] Jiao J, D'haeseleer P, Dill BD, Shah M, VerBerkmoes NC, Hettich RL, Banfield JF, Thelen MP. Identification of biofilm matrix-associated proteins from an acid mine drainage microbial community. Appl Env Microbiol; 2011;**77**(15):5230–7. doi: 10.1128/AEM.03005-10.

Chapter 8

Metabolomics as a Tool in Agriculture

Emmanuel Ibarra-Estrada,

Ramón Marcos Soto-Hernández and

Mariana Palma-Tenango

Additional information is available at the end of the chapter

http://dx.doi.org/10.5772/66485

Abstract

Metabolomics is a study through can be obtain a better understanding of the complexity of the biological systems, through the chemical composition and relations with the physiology of the plant. The literature describes a lot of information on this support in areas of medicinal plants, chemosystematics, adulteration of plants, etc., but it is scarce in agriculture. At present, agriculture plays a crucial role in human beings. The demand of foods has increased due to the continuous increase in the population, and this requires an increase in the production of crops, besides the crops being affected by the climatic change, attack of pest and diseases and resistance to a conventional agrochemicals. At present, scientists are doing some practices or studies of genetic improvement of crops to increase their production and avoid the problems pointed out. It is an important part of the genomic studies; the results could be the basis for a genetic improvement based on the chemical composition of the crops, and in the metabolomics studies represent a crucial role in their quality for human consumption. The aim of the chapter is to review the literature from 10 years behind emphasizing the importance of metabolomics in crops of economic and feed value.

Keywords: agriculture, metabolomics, crops

1. Introduction

Plants are a large source of natural products with a large chemical diversity and unimaginable properties; therefore, the interest in their study is increasing. Nowadays, the interest in agricultural crops has increased due to their economic importance and their projected importance in the next years in relation to world food security.

The usage on the plant species destined to agriculture is varied, from traditional foods to those with some desirable traits, such as nutritional value and the industrial products derived from them such as polymers, fibres, latex, industrial oils and packaging materials, in addition to basic chemical building blocks and fuels [1].

The objectives of the use of metabolomics tools in agriculture are to know the biochemistry and the functions of the species involved to apply that knowledge to food and environmental security; to use their potentials as tools in the improvement of nutrition, diets and health [2]; and to use them for genetic improvement in plants based on chemical composition, that is, taking into account some notable trait.

Nowadays, agriculture plays a crucial role in human feeding. Food demand has increased considerably due to the continuous increase of population centres, and thus, the demands on the production and diversity of crops, especially basic crops, are increasing. To this problem, we can add the noxious effects of climate change on crops, the attack of pests and diseases, as well as the emergence of new diseases and the development of resistance by them towards commonly used pesticides. Currently, trials or studies of genetic improvement of crops are being carried out to increase the quantity and quality of yield, avoid damages caused by pest and diseases and develop resistance to several factors, especially environmental ones. There are some crops that are considered as basic or important to world food, such as rice, maize, potato, avocado, tomato and citrus fruits, among others; these crops are placing themselves among priority crops because they support human consumption and thus have been the subject of the aforementioned studies or trials.

2. Metabolomics

Metabolomics represents a field of study with which we can gain a better understanding of the complexity of biological systems. It deals with the identification and quantification of the metabolites present in such systems [3, 4] with molecular weights less than 1500 Da [5], although the range could be occasionally wider (30–3000 Da) [6]. The group of small molecules of metabolites that are in a cell, in an organ or organism, is called metabolome [6, 7]. It is made up of a large variety of molecules such as peptides, amino acids, nucleic acids, carbohydrates, organic acids, vitamins, flavonoids, polyphenols, alkaloids, minerals or any other chemical compound that is used, metabolized or synthesized by a cell or by a given organism [7]. The importance of the metabolites resides in that they are an essential part of the behaviour of the individual that contains them; these compounds are the final products of the regulatory processes of the cell, and their presence represents the response of biological systems to environmental or genetic changes. It is due to this that metabolomics is considered as the link between genotype and phenotype [8, 9]. In the last years, metabolomics has managed to position itself as an area of research that is essential and complementary to proteomic, genomic and transcriptomic studies.

Metabolomics is a very important complement of genomic studies, and its results could be the basis of genetic improvement based on the chemical composition of crops, be it in nutritional or functional aspects, or in the participation of some chemical compounds in the resistance of some plant species to the factors that have already been mentioned.

It is important to know the current state of metabolomics studies of the most important crops in the world to use it as the basis for future studies destined to improve their production, or to know the diversity of the chemical composition that we have available, its benefits and its possible uses. Moreover, it would be important to know the advantages and disadvantages of using current analytical techniques, in addition to their current state and possible improvements in future analyses.

3. Extraction methods in metabolomics

There are numerous extraction methods used to extract and isolate the compounds of interest, but is important to keep in mind to use a simple method, robust, low consume time, repeatable and cheap. The selected solvent should extract diverse groups of metabolites. Traditional extraction methods as percolation, maceration, Soxhlet extraction, steam distillation or hydrodistillation are convenient to extract a broad class of metabolites; they are cheap, simple and

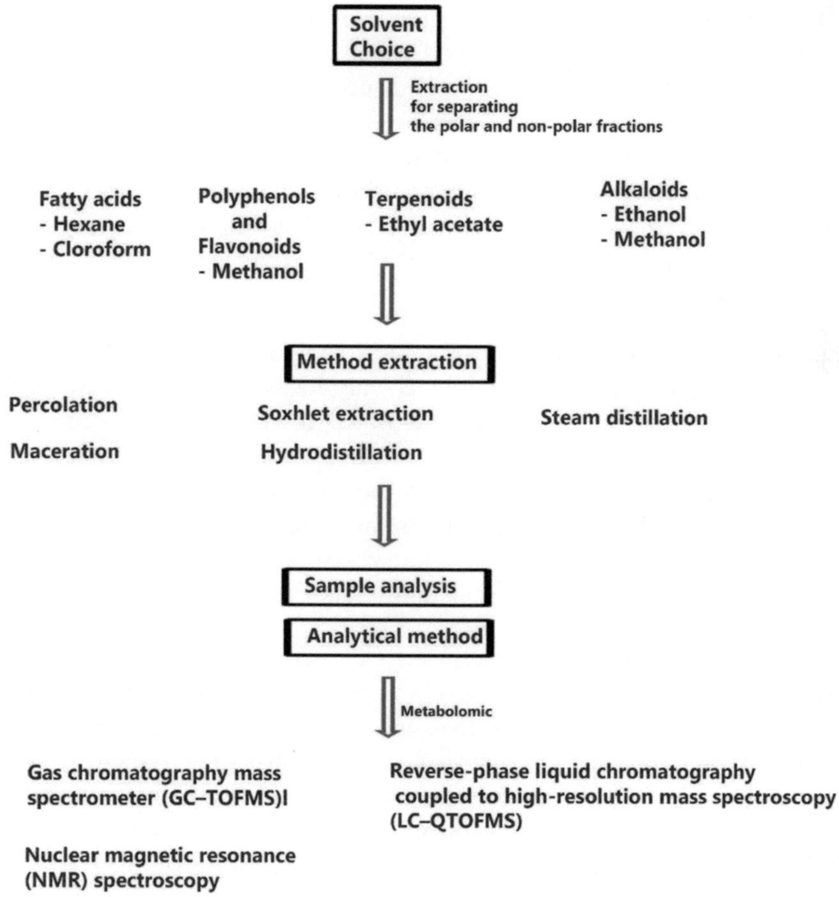

Figure 1. Extraction process and analysis of a sample. Modified from Ref. [92].

repeatable and can be used for raw plant extraction; and the amounts to be processed depend on the source of plant material, and the amounts to be processed can be since a few grams to a higher amounts, but depends of the source of plant material, but they are time consuming (**Figure 1**). At present, they can complemented with modern techniques as ultrasonication (sonochemistry), microwaves, supercritical fluid extraction or accelerated solvent extraction; they are simple, repeatable and mainly the extraction time is short, but the cost of these equipments is high. It is also important to consider the effect of temperature provided by the selected method, because some components of the sample can be decomposed if the temperature is high. Roesnner and Dias [10] described a recent review where they notice all the details that should be consider to get a good result with the extraction and isolation of the compounds of interest. Once the sample is extracted properly, is ready to submit for the analysis on GC-MS, LC-MS, NMR or MS.

4. Analytical techniques

Today we have a wide variety of analytical techniques that are getting better and better, and which help us to obtain reliable data about the behaviour of a plant species, or about its response to diverse environmental factors, both biotic and abiotic. During the last years, there has been a boom of genetic studies of many plant species, and the generated information can be used in a wide range of applications, for example, genetic improvement. A very important aspect that could help us to improve the understanding and to enrich the information of such studies would be the inclusion of metabolomic studies. Metabolomics has been considered as the link between genotype and phenotype, that is, with this kind of studies, gene expression would be better understood, and hence, the behaviour of the plant species in the face of the aforementioned factors.

An essential aspect of the metabolomic study is the qualitative and quantitative analysis of metabolites, which allows us to know the biochemical state of an organism; that information can be used to monitor and evaluate gene function and the multiple responses of the organism to the conditions where it develops [8]. In spite of the technological advances, it is practically impossible to determine the entire composition of even a single cell in a single study or with a single analytical technique [9]; in order to do that, coupled techniques have been developed, such as high-performance liquid chromatography (HPLC) and gas chromatography (GC) coupled with mass spectrometry (MS).

The analyses that have already been mentioned are carried out using analytical techniques specialized in separation, identification and quantification. Such techniques must have high resolution, be very precise and very sensitive, and be able to analyse a wide variety of compounds of different chemical nature and origin because the structural complexity of many molecules makes their study difficult. Several analytical technologies can be used, depending on the chemical nature of the compounds. Some of them are nuclear magnetic resonance (NMR), GC and HPLC coupled with MS, as well as capillary electrophoresis coupled with MS [7].

It is known that a single analytical technique is not enough to visualize whole metabolome [11], and thus, it is necessary and important to carry out separate studies using different techniques, or using the aforementioned coupled systems.

NMR has several uses in metabolomics which can be applied or adapted to agriculture. For example, in quality control, chemotaxonomy (classification and characterization), analysis of genetically modified plants and the study of interactions with other organisms and the environment, besides the study of diseases in humans. Currently, the study of the metabolome based on NMR is accepted as an efficient analytical tool to study biological systems [6].

The main advantages of the NMR over the rest of the analytical techniques are that it can detect a wide range of chemical compounds of different nature; quantification does not pose any problem; it is highly reproducible and metabolite identification is simple [12]. Perhaps the most important benefit is that the method is quick and simple, and the damage of the existing compounds during the preparation of the extracts is minimal. However, the technique is not very sensitive [13, 14].

As it was already mentioned, chemotaxonomy is one of the main applications of NMR, since through the metabolomics profiles obtained by this technique it is possible to classify and identify plants and their derived preparations. There are several examples of the application of NMR in chemotaxonomy. The classification of Ilex species [15], *Cannabis sativa* [16], *Ephedra* [17], and the metabolomic differentiation and classification of species of *Verbascum* [18], discrimination of commercial preparations of *Matricaria* [19] and commercial samples of catuaba [20], among others.

5. Applications of metabolomics in agriculture

Specifically in agriculture, metabolite content is related to developmental and differentiation processes, fruit maturation processes, resistance to adverse environmental factors, stress-related problems and pathogen attack, among others.

The wide range of compounds is analysed through several analytical techniques; some of which are specific to certain compounds. Liquid chromatography coupled with mass spectrometry is a technique that can be used to analyse a wide range of compounds such as vitamins (hydrophilic and hydrophobic), coenzymes, phenylpropanoids, polyketides, terpenoids, amino acids and amines, lipids, carbohydrates, phenolic compounds and alkaloids, among others. With GC-MS, fewer compounds can be analysed due to the type of compounds that this technique can detect; although by using derivatization reactions, the number of metabolites detected by this technique increases considerably. With this technique, it is easy to analyse essential oils and fatty acids as well as terpenoids, alkaloids, monosaccharides and steroids, among others. Capillary electrophoresis, for its part, helps in the detection of oligosaccharides, hydrophobic vitamins, coenzymes, prosthetic groups, nitrogenous bases, nucleotides and nucleosides, among others [21].

Crop or plant species	Compound/analytical technique	Topic	References
Avocado	Mannoheptulose carbohydrate)/LC-MS	Development and growth	[86, 87, 91]
	Perseitol (carbohydrate) /LC-MS	Development and growth	[85, 86, 90]
	Stearic, palmitic, oleic, Linoleic, linolenic and palmitoleic acids (fatty acids), carbohydrates /FID-GC, LC	Development of rapid method	[88]
	Persenone A and B/HPLC, NMR	Targeted metabolomics	[53]
	Cyanidin 3-O-glucoside (anthocyanin)/HPLC	Fruit ripening	[84]
	Sugars, protein, oil/LC	Fruit ripening	[54]
Tomato	1-Methyl-tryptophan/UHLC-MS	Plant-pathogen interaction	[50]
	Flavonoids/GC-MS, LC-MX	Metabolic engineering	[51]
	Sugars, amino acids, organic acids/NMR, GC-TOF-MS	Growth and development	[49]
	Organic acids/NMR, GC-MS	Metabolic engineering	[89]
	Enzymes/GC-MS	Fruit development in transgenic plants	[90]
Maize	Amino acids, carbohydrates, organic acids/NMR	Salt stress	[27]
	Total nitrogen, protein/MS-MS	Development and growth	[85]
	Protein/MALDI-TOF-MS	Metabolic changes under phosphorus deficit	[26]
Potato	Glycoalkaloids, fructans/GC-TOF-MS, LC-MS	Genetically modified plants	[37]
	Fatty acids, amino acids, organic acids/NMR, HPLC-UV	Genetic modifications to metabolic pathways	[39]
	Sugars, amino acids, organic acids/FIE-MS, GC-MS	Total composition analysis and quality trait	[36]
	Fatty acids, organic acids, alkaloids and amino acids	Plant-pathogen interaction	[40]
	Amino acids, organic acids, sugars and sugar alcohols	Phytochemical diversity	[34]

Table 1. Metabolomics studies in agricultural crops.

Currently, there are many examples of metabolomic studies that have been applied to agriculture with various purposes (**Table 1**). Some of the main objectives of metabolomic studies in agriculture are as follows: know the metabolic responses towards any type of stress; generate metabolic profiles for genetic mapping and generate metabolic profiles to study the impact of heredity. Generate metabolic profiles to determine the impact of geographic location and season, study the phenotype of natural variations of certain plant species, evaluate transgenic varieties, study metabolic variations of different cultivars and carry out functional characterizations (functional genomics). Analyse metabolic changes during growth, development and differentiation, elucidate biosynthetic pathways, analyse population differentiation, carry out chemotaxonomic analyses, carry out nontargeted studies and characterize cultivars [21].

Nowadays, human world food depends on certain crops that are indispensable due to their nutritional value, and the access that we have to them.

Maize (*Zea mays* L.) has been subject of numerous metabolomic studies, many of which are related to genomic studies. For example, it has been obtained the metabolomic profile of its leaves to carry out genetic mapping using GC-MS [22]. On the other hand, the genetic basis of the metabolic diversity in its grains was analysed in a study that involved 702 genotypes collected from different geographic zones and growing conditions; in that study, 983 metabolites were quantified [23]. Another study that included the analysis of genetic background (heredity), different geographic locations and environmental conditions through GC-MS was the one carried out by Röhlig et al. [24]; in that study, they could distinguish the chemical composition of the genotypes under analysis, which helped to determine notable chemical characteristics that varied depending on the region of origin.

The development of maize, as well as of other crops, under ideal nutritional conditions is paramount in order to obtain good yield and quality; hence, the importance of carrying out studies that involve nutrient analysis. The effect of nitrogen (N) deficiency on the development of maize has been studied using metabolomic studies [25], and the effect of phosphorous (P) deficiency has been analysed through MALDI-TOF MS [26]. Stress caused by any factor limits the growth and development of any plant species. The effect of salt stress on the growth of maize seedlings was studied through ^{1}H-NMR; the metabolic profile obtained in the study can be used to improve the growing conditions [27].

Rice (*Oryza sativa* L.) is one of the most important crops in the world, especially in Asia. Several studies have been carried out in this crop, where the main objective has been to characterize cultivars through the metabolic profiles obtained from several plant structures, mainly the leaf, using capillary electrophoresis coupled with mass spectrometry [28]. This crop has also been subject of studies where genetic and metabolic studies are combined using techniques such as GC-FID, GC-MS and GC-TOF-MS, in what is called functional genomics [29, 30]. Naturally, in modern times, studies on transgenic individuals could not be absent, where their metabolic profiles have been determined [31].

Potato (*Solanum tuberosum* L.) is a crop that has been widely studied in several areas, among which is metabolomics. In such studies, the objectives have ranged from the elucidation of the metabolic profile to know the chemical characteristics of the tubers through GC-MS [32, 33] and the characterization of some cultivars [34, 35], to the characterization of the relationship between composition and quality traits [36].

As it was mentioned in maize and rice, there have been analyses in transgenic potato crops to know their metabolic profile; such studies have been carried out in tubers through techniques such as GC-MS [37, 38] proton NMR and HPLC-UV [39].

One of the main limitations in yield and quality of crops is the effects caused by pathogens. It is important to study the interaction between pathogens and crops to gain a better understanding of its effects. In this crop, the interaction with *Rhizoctonia solani* has been studied using FT-ICR/MS and GC-EI/MS [40].

Metabolomics in agriculture can be used to obtain a chemotaxonomic classification, where differences and similarities can be distinguished between related species or cultivars, as it was the case between genetically modified potato crops and conventional potato crops, where the study was carried out using GC TOF-MS and FIE-MS [37].

Species in the family Solanaceae, such as potato and tomato (*Solanum lycopersicum* L.), are among the most important crops for agriculture and food and thus are subject of different types of studies. With respect to metabolomic studies, metabolomic profiles of tomato have been obtained for genetic mapping of fruits and leaves of both, conventional and wild genotypes [41, 42]. It is known that tomato has a large number of chemical compounds with important uses. This chemical variability has been elucidated through metabolomic studies. An important group of compounds that have been studied is the volatiles from fruit, peel and pulp, which were analysed through UPLC-QTOF-MS [43, 44]. The profile of carotenoid compounds from the fruit has also been characterized by proton NMR [45]. Changes in chemical composition during growth, development and differentiation are very important to determine the conditions where these events are occurring. It has been obtained chemical profiles from tomato plants in several stages of development through LC-MS [46]. Something similar was carried out by Moco et al. [47], where they obtained databases from the metabolome from the fruit using LC-MS and ^{1}H-NMR. The simple characterization of fruits from different cultivars was carried out through MALDI/TOF-MS [48].

During the day, any plant species has changes in its chemical composition due to the action or effect of the environment, something that [49] determined when detected 70 metabolites in tomato leaves and 60 in the fruit using NMR and MS. Also, with the aid of metabolomics, it is possible to identify particular compounds with some specific function, important to the survival of a species. With respect to this, 1-methyltryptophan was identified as the metabolite involved in the response of the plant to *Botrytis cinerea* and *Pseudomonas syringae* [50].

On the other hand, it can be altered the levels and composition of flavonoids in tomato fruits through the modification of biosynthetic pathways using regulatory and structural genes [51].

Throughout time, about 95% of the genetic and chemical diversity of tomato has been lost owing to its domestication; therefore, it is important to know the entire metabolic profiles of both commercial and surviving wild tomatoes to determine their chemical and genetic variability and use that information as the foundation to obtain better genotypes or varieties [52].

One of the most important fruit trees, economically speaking, due to its particular characteristics as flavour, texture and chemical composition, is the avocado (*Persea americana* Miller).

It has also been subject of different studies where its main chemical traits have been determined. It is a species with important properties and benefits to health, based on several chemical compounds.

The study of the chemical compounds present in avocado is very important, and it will help us to lay the foundations to use the chemical information of this fruit tree, for example, in genetic improvement or it can be used to increase the interest in those varieties or races that are not widely consumed and find new applications to them. In this approach, the chemical composition of avocado is addressed using techniques and strategies that combine the identification and quantification of cellular metabolites through sophisticated analytical techniques, and the use of statistical and multivariate methods to analyse and interpret data. In the development of analytical methods throughout time, there have been important technological advances, which have resulted in improvements in the way biological systems are seen, analysed and interpreted [13].

In the last years, the content of acetogenins in avocado has attracted attention. These long-chain fatty acid derivatives have important medicinal properties; they are considered anticancer agents. The profile of acetogenins in the peel, seed and pulp of avocado fruits has been determined to obtain a chemotaxonomic model using a linear discriminant analysis [53]. Also, changes of sugars, total protein and oil in 'Hass' avocado have been determined using GC-FID and LC [54]. On the other hand, important proteins from the pulp have been identified through the use of nano-LC-MS/MS [55]. The use of coupled analytical systems has helped to get better resolution, higher sensitivity, higher speed of analysis and wider diversity of applications. A study in avocado was carried out to investigate fruit maturation through its metabolomic profile using GC-APCI-TOF-MS. Such technique showed that it is a valuable and powerful tool to improve the understanding of the process of fruit maturation [56].

Through the years, the metabolomic knowledge has been applied to agriculture through various planned objectives. For example, some studies have been carried out to elucidate the biosynthetic pathways that produce metabolites in herbs in order to discover the mechanisms responsible for the evolution of these pathways and to understand the function of a given natural product within the physiology of the plant where it is found [2]. Such studies have been carried out in ginger (*Zingiber officinale* Roscoe) [57], turmeric (*Curcuma longa* L.) [58] and basil (*Ocimum basilicum* L.) [59]. These plants have an enormous potential for the development of products with some application, mainly industrial or in medicinal.

A metabolomic study, whatever it may be, is a process with some key stages to know the real status of a plant. Such stages are collection and extraction of the sample, and the analysis, identification, and quantification of the chemical compounds of interest, according to the objective of study. Another case is the analysis of the extracts of *Pisum sativum* L. to define the impact of the environment and the genetic diversity based on the metabolic profile [60, 61].

Several metabolic studies have been carried out in cultivable plant species, where all the tissues have been used and all the analytical techniques employed. Within the cereal group, the leaves and roots of barley (*Hordeum vulgare* L.) have been studied to analyse the metabolic responses to salt in two cultivars using GC-MS [62]. In oat, the metabolic profile of leaves was obtained using ^{1}H-NMR to evaluate metabolic variation in European cultivars [63]. Several studies have been carried out also in legumes. For example, in pea (*Pisum sativum* L.), the metabolic fingerprint of transgenic varieties was obtained from the leaves using ^{1}H-NMR [61]. In *Medicago truncatula* Gaertn, the functional characterization of glycosyltransferases of terpenes in cell culture was obtained using HPLC-MS and also the metabolomic profile for functional genomics in cell culture [64, 65]. In the seeds of *Vigna radiata* (L.) R. Wilczek, the metabolic changes that occur during sprouting were investigated through GC-MS [66].

The tobacco (*Nicotiana tabacum L*) has been subject of study in functional genomics to study the changes induced by jasmonates in the biosynthesis of metabolites in a cell suspension culture, using GC-MS [67]. Also in this family, the metabolic profile from the fruit of chili (*Capsicum* sp.) was determined to study its diversity, using HPLC-PDA/LC-PDA-QTOF-MS-MS [68].

Fruit trees have also been studied to obtain their metabolic profiles for several purposes. In grape berries (*Vitis vinifera* L.), the metabolic variation in several cultivars was evaluated using ^{1}H-NMR [69]. Also, the metabolic profile of raspberry fruits (*Rubus idaeus* L.) was characterized to identify beneficial compounds using LC-MS [70]. In strawberry (*Fragaria × ananassa* hort.), metabolic studies in fruits and flowers were carried out to know the chemical composition at several developmental stages [71, 72]. Finally, in musk-melon (*Cucumis melo* L.), the metabolic profile of the fruit was obtained using GC-MS [73]. Medicinal plants are also subject of study to find medicinal chemical compounds. In *Catharanthus roseus* (L.) G. Don, there has been hundreds of studies to understand and improve the biosynthetic pathway of indole alkaloids in cell culture; many of the studies have used LC-MS and ^{1}H-NMR.

The analytical techniques have many advantages for metabolomic analyses; however, their application is not universal. It is important to know the limitations and the processes that are being developed to improve them.

Metabolomic approaches have been divided in two groups, targeted and untargeted metabolomics. In principle, both approaches can be applied in all analytical platforms as LC-MS and GC-MS. The main difference between them is the identification of the analytical signals. In the first one, the metabolites under study are known. In the second one, no specific metabolites are chosen and all the signals detected are taken in account. Clearly, both techniques have advantages and disadvantages. The main benefit of untargeted approach is a more holistic view of the behaviour of metabolite composition with the chance of low probability of missing key metabolites. If it already known what specific metabolites are key for the research, an optimized targeted approach could be most successful. Through the use of internal standards, analysis can be undertaken in a quantitative or semi-quantitative form.

These tools have been useful in pre-harvest and postharvest issues of food constituents and are related with food safety and quality control. In fruits as mango prone to a pre-harvest fungal disease or the postharvest contamination of onions have been assessed by GC-MS, in which the sample is trapped by head space and then submitted for the GC-MS analysis. In fact, metabolomics techniques may find their greatest application in the food industry in monitoring quality control of different batches. It is described that fruit juice adulteration is quite common; it is not easy to detect the adulteration, but the LC-MS has helped to control the adulteration. The distinction between fresh squeezed juices and those come from pulp washes also has been determined by ^{1}H NMR, with a high accuracy of the method [4]. Recently, Canela et al. [83] described a review on foodomics imaging by mass spectrometry and magnetic resonance (IMS and MRI). They pointed out that these tools have advantages over fluorescent microscopy or immunochemistry to localization and chemical identity of small molecules. These can determine the presence of many compounds in a single multi-detecting measurement. In the review described some examples where these techniques can support the visualization of the compounds present in a plant tissue used for evaluation of quality control.

In general, there are many uses of metabolomics in agriculture, and there are no limitations in the study of the multiple species. It is possible to work with any plant structure, with any crop, for almost any purpose, and the array of techniques available makes it easy to carry out these studies. In the end, the results of metabolomics from agriculture will have an impact on other areas of study, such as medicine, food quality control, nutrition, and genetic improvement, among others. Finally, an essential part of metabolomic studies is the statistical analysis and bioinformatics resources used to interpret the results, which are important to take into account in a wide review of metabolomics.

6. Plant-pathogen interaction

Plants are always influenced by different factors, mainly environmental ones; those factors force them to adapt and change parts of their functioning to protect themselves, in most cases. One of the main interactions that exist in plants is that with micro-organisms, which also cause changes in the physiology and development of plants, like the environment does. Normally, there is an adverse effect to the plant, but in some occasions, both obtain a benefit, in what is called symbiosis. Pathogen attacks can lead to yield losses although occasionally the micro-organisms may help the plant to improve nutrient uptake. The most characteristic example is the interaction of nitrogen-fixing bacteria with legumes.

When the interaction between a plant and a pathogen occurs, one of the changes that happen in the plant is the production of several kinds of compounds that act as a type of defence as attractants, repellents, feeding inhibitors or the production of chemical compounds that are beneficial to human health.

Metabolomics, through its variants, can help to determine the physiological and biochemical changes that happen during the interaction and provide a general overview of the whole system.

Primary and secondary metabolites are a key in the response of the plant towards pathogen attacks. There are many modifications that could happen during such interaction, such as molecular and physiological modifications. The modifications range from changes in primary metabolism, where basic processes like photosynthesis can be affected, modifications in the cell wall and in some organs of the plant, to the production of secondary metabolites that can be toxic or trigger defensive signals in the plant. Depending on the type of interaction, resistance, tolerance or susceptibility could occur [74].

One example is the plant-fungus interaction, where a study was carried out using mass spectrometry using electrospray ionization to detect changes in the levels of lipids and hormones. In this study, the researchers had predicted that those molecules were involved in the interaction between *Brachypodium distachyon* and *Magnaporthe grisea*. A variation in the level of phospholipids was detected, which was the main response of the plant to the attack of the fungus [75]. One advantage of the use of metabolomics is that targeted and nontargeted studies can be performed at the same time, as in the case of the interaction between *Lupinus angustifolius* with the fungus *Colletotrichum lupini*. The response of this interaction was characterized focusing on the metabolites of the cuticle using GC-MS (nontargeted study) and on the flavonoids using LC-MS (targeted study). In this case, it was found that there was a higher variation in the kinds of flavonoids when the interaction took place and that variation was more marked in the extract from the plant than in the extract from the fungus [76].

On the other hand, a metabolomic study of the plant-pathogen interaction can help to elucidate the genetic mechanisms originated during resistance. Through a nontargeted study using GC-MS, the resistance of sunflower (*Helianthus annuus*) to the necrotrophic pathogen *Sclerotinia sclerotiorum* was characterized; 63 metabolites were found, including sugars, organic acids, amino acids and secondary metabolites, such as chlorogenic acid, which are associated with tolerant phenotypes [77].

Many studies about plant-pathogen interactions have focused on relating chemical compounds with diseases using bacteria, fungi, oomycetes and even viruses, interacting with plants such as Arabidopsis, tobacco, sunflower, barley, rice, potato and grapevine, where changes have been detected in both primary and secondary metabolism [74]. Most of the studies that have been mentioned have used mass spectrometry (coupled with LC or GC) as the analytical technique that helps to characterize the metabolomic response of the interaction, but there are studies based on NMR to achieve the characterization. In a study of the interaction between tobacco (*Nicotiana tabacum* L.) and the tobacco mosaic virus, the alteration of the metabolic pathways in the leaves was proposed after they were infected with the virus. Healthy and diseased leaves were compared through 1D and 2D NMR [78].

NMR has also been used to characterize the compounds involved in the resistance of host plants to the western flower thrips (*Frankliniella occidentalis*). For this study, three plants of different types were used: Senecio (wild), chrysanthemum (ornamental) and tomato (cultivated). A resistant Senecio hybrid was evaluated because it showed significantly higher amounts of pyrrolizidine alkaloids (which are involved in plant chemical defences against herbivores in general), as well as some flavonoids. In the case of the resistant chrysanthemum, high amounts of chlorogenic acid and ferulic acid were found which, as phenolic compounds, are expressed during plant defence. Cultivated tomato was the most susceptible towards pathogen attack, but the resistant hybrid of this crop showed high levels of acyl sugars, which confer some protection against the attack of pathogens [79].

There are many examples of the application of metabolomics in plant-pathogen interactions that help us to know and interpret this interrelation through the physiological and biochemical changes that take place. For this, advanced analytical platforms of high sensitivity are used; these platforms allow the characterization of almost any molecule expressed during the interaction. Any kind of plant involved in any type of attack or interaction can be studied, and valuable information can be obtained that will help us to understand the reaction of the plant or both. From this type of interactions, and through applied analytical techniques, information about the best way to exploit such interaction can be obtained, in case one wants to obtain or increase chemical compounds with some use, or if one has interest in other aspects of life, such as health.

7. Future perspectives of metabolomics

Over the years, the field of metabolomics has gained interest in several disciplines such as functional genomics, biological and integrative systems, pharmacogenomics, and the discovery of biomarkers to predict diseases, diagnostics and therapy monitoring [80], besides the area of study of agriculture [2].

Modern challenges for metabolomics are diverse in all the fields mentioned, but they are particularly relevant in the discovery of biomarkers [82], especially in the field of diseases, because their detection, monitoring and treatment are important. One of the problems that researchers face these days is the difficulty in the identification and quantification of several chemical compounds at the same time in a reliable and ideal way, since their number can be huge, and some metabolites, which may be undetectable due to their low concentrations, may be relevant to some function or application. Nowadays, new analytical techniques are being developed or improved to solve these problems, widening the range of detectable metabolites based on their structural characteristics, and making the methods more sensitive, so they can detect very low concentrations of compounds.

Nowadays, the information obtained from analysed samples may not be correctly interpreted, or it may offer much more information than what is obtained, but some limitations related to the design of the experiment could have impeded the correct interpretation and use of that information.

A future challenge will be to improve this type of situations to use the information better and find possible applications for it. The validity of a metabolomic study is affected by the sizes of unbalanced samples (this aspect has to do with the design of the experiment), especially in studies with humans, mainly when statistical methods are used to interpret the data. This is mentioned because in some studies that have already been carried out, the number of control and diseased individuals is not associated, that is, it is not balanced [81], so it is necessary to balance these cases of study.

One of the challenges of metabolomics is to participate in more fields of study which could be waiting for this type of analyses; although the most important ones are already included, such as disease detection and health in general, studies have already started in fields such as evolution, chemotaxonomy, agriculture, ecology and food quality control, among others.

Nowadays, there are many databases related to chemical compounds, their identification and their structural elucidation, and probably over the years these databases will increase significantly due to the increasing number of studies in this area; therefore, we will have access to a huge number of metabolites, their properties, their possible health benefits and other properties.

Certainly, with the passing of the years, the number of studies related to the functionality of what is already sequenced will increase, since in this field there are still some delays. This is the main challenge of metabolomics for the next years, and it will be conquered by integrating the study of several fields focused on the same objective, or by integrating the latter.

The complete understanding of the function of the cell system and the deciphering of gene function will arrive with time, since metabolomics is integrated with genomics, transcriptomics and proteomics; thus, an integral work will give the result that everyone wants in this branch of knowledge, which is the complete understanding of cell function [81].

Chemical studies in plants, from their origins, have been based on their traditional use and knowledge, since people used them to get some benefit, but without really knowing what caused such effect. Therefore, something that must be addressed is to make the composition and the beneficial effects of the compounds available to the general public in an easy-to-understand way.

Metabolomic studies in plants related to natural products will increase simply because they are everywhere, they can be used for everything, and every day new applications are found for them; most of these applications will benefit humans.

It has mentioned that metabolomics is not a goal in itself, but a tool to improve our understanding about the metabolism and biochemistry of the organisms [6]. Therefore, among all, this must be the most important future perspective: to know and completely understand metabolism and cell function.

8. Conclusion

Metabolomics is a relatively new field. During the last years, this discipline has been growing because diverse applications have been found for it, and different analytical techniques have been developed and improved; this has allowed an easier interpretation and analysis of the results. The inclusion of a wide variety of crops in this type of studies is paramount, because it is necessary to know the qualities that they have, and take from them the most important traits with the aim of developing an application that benefits food, health or industry.

The field of study of agriculture, with respect to metabolomic aspects, will keep growing, because in the next years, there will be challenges to ensure world food sovereignty. Currently, most crops and their diversity are at risk; therefore, it is necessary to carry out actions focused on their conservation, rescue and rational exploitation.

The range of analytical techniques implemented in metabolomics allows us to be a step ahead in the analysis of extracts or chemical compounds, through which new uses or applications for the plant species studied can be found, or strengthen those already existing, in addition to the development of improvement programmes based on distinctive chemical traits.

Author details

Emmanuel Ibarra-Estrada[1], Ramón Marcos Soto-Hernández[2]* and Mariana Palma-Tenango[2]

*Address all correspondence to: msoto@colpos.mx

1 Ministry of Agriculture, México City, México

2 Postgraduate Studies College, Campus Montecillo, Texcoco, México

References

[1] Abbas C A, Cheryan M: Biorefinery opportunities. Applied Biochemistry and Biotech nology. 2002;**98**:1147.

[2] Dixon R A, Gang D R, Charlton A J, Fiehn O, Kuiper H A, Reynolds T L, Tjeerdema R S, Jeffery E H, German J B, Ridley W P, Seiber J N: Applications of metabolomics in agriculture. Journal of Agricultural and Food Chemistry. 2006;**54**:8984–8994.

[3] Verpoorte R, Choi Y H, Kim H K: NMR-based metabolomics at work in phytochemistry. Phytochemistry Reviews. 2007;**6**:3–14.

[4] Cevallos-Cevallos J M, Reyes-De-Corcuera J I, Etxeberria E, Danyluk M D, Rodrick G E: Metabolomic analysis in food science: a review. Trends in Food Science & Technology. 2009;**20**:557–566.

[5] German J B, Hammock B D, Watkins S M. Metabolomics: building on a century of biochemistry to guide human health. Metabolomics. 2005;**1**:3–9.

[6] Kim H K, Choi Y H, Verpoorte R: NMR-based plant metabolomics: where do we stand, where do we go? Trends in Biotechnology. 2011;**26**:267–275.

[7] Wishart D S: Metabolomics: applications to food science and nutrition research. Trends in Food Science & Technology. 2008;**19**:482–493.

[8] Fiehn O: The link between genotypes and phenotypes. Plant Molecular Biology. 2002;**48**48:155–171.

[9] Kobayashi A: Plant metabolomics: potential for practical operation. Journal of Bioscience and Bioengineering. 2005;**100**:347–354.

[10] Roesnner U, Dias D A: Plant tissue extraction for metabolomics. In: Springer-Verlag, editor. Metabolomics Tools for Natural Products Discovery: Methods and Protocols in Molecular Biology. Springer-Verlag, New York. 2013. p. 21–28.

[11] Hall R, Beale M, Fiehn O, Hardy N, Sumner L, Bino R: Plant metabolomics as the missing link in functional genomics strategies. Plant Cell. 2002;**14**:1437–1440.

[12] Verpoorte R, Choi Y H, Mustafa N R, Kim H K: Metabolomics: back to basics. Phytochemistry Reviews. 2008;**7**:525–537.

[13] Sumner L W, Mendes P, Dixon R A: Plant metabolomics: large-scale phytochemistry in the functional genomics era. Phytochemistry. 2003;**62**:817–836.

[14] Kim H K, Choi Y H, Verpoorte R: NMR-based metabolomic analysis of plants. Nature Protocols. 2010;**5**:536–549.

[15] Kim H K, Saifullah, Khan S, Wilson E G, Prat Kricun S D, Meissner A, Goraler S, Deelder A M, Choi Y H, Verpoorte R: Metabolic classification of South American Ilex species by NMR-bases metabolomics. Phytochemistry. 2010;**71**:773–784.

[16] Choi Y H, Kim H K, Hazekamp A, Erkelens C, Lefeber A W M, Verpoorte R. Metabolomic differentiation of *Cannabis sativa* cultivars using ^{1}H-NMR spectroscopy and principal component analysis. Journal of Natural Products. 2004;**67**:953–957.

[17] Kim H K, Choi Y H, Erkelens C, Lefeber A W M, Verpoorte R: Metabolic fingerprinting of Ephedra species using ^{1}H-NMR spectroscopy and principal component analysis. Chemical and Pharmaceutical Bulletin. 2005;**53**:105–109.

[18] Georgiev M I, Ali K, Alipieva K, Verpoorte R, Choi Y H: Metabolic differentiations and classification of Verbascum species by NMR-based metabolomics. Phytochemistry. 2011;**72**:2045–2051.

[19] Bailey N J C, Sampson J, Hylands P J, Nicholson J K, Holmes E: Multicomponent metabolic classification of commercial feverfew preparations via high-field ^{1}H-NMR spectroscopy and chemometrics. Planta Medica. 2002;**68**:734–738. doi:10.1055/s-2002-33793

[20] Daolio C, Beltrame F L, Ferreira A G, Cass Q B, Cortez D A G, Ferreira M C: Classification of commercial catuaba samples by NMR, HPLC and chemometrics. Phytochemical Analysis. 2008;**19**:218–228.

[21] Carreno-Quintero N, Bouwmeester H J, Keurentjes J J B: Genetic analysis of metabolome-phenotype interactions: from model to crop species. Trends in Genetics. 2013;**29**:41–50.

[22] Riedelsheimer C, Lisec J, Czedik-Eysenberg A, Sulpice R, Flis A, Grieder C, Altmann T, Stitt M, Willmitzer L, Melchinger A E: Genome-wide association mapping of leaf metabolic profiles for dissecting complex traits in maize. Proceedings of the National Academy of Sciences of the United States of America. 2012;**109**:8872–8877.

[23] Wen W, Li D, Li X, Gao Y, Li W, Li H, Liu J, Liu H, Chen W, Luo J, Yan J. Metabolome-based genome-wide association study of maize kernel leads to novel biochemical insights. Nature Communications. 2014;**5**:1–10.

[24] Röhlig R M, Eder J, Engel K H: Metabolite profiling of maize grain: differentiation due to genetics and environment. Metabolomics. 2009;**5**:459–477.

[25] Simons M, Saha R, Guillard L, Clément G, Armengaud P, Cañas R, Maranas C D, Lea P J, Hirel B: Nitrogen-use efficiency in maize (*Zea mays* L.): from 'omics' studies to metabolic modelling. Journal of Experimental Botany. 2014;**65**:5657–5671.

[26] Li K, Xu C, Zhang K, Yang A, Zhang J: Proteomic analysis of roots growth and metabolic changes under phosphorus deficit in maize (*Zea mays* L.) plants. Proteomics. 2007;**7**:1501–1512.

[27] Gavaghan C L, Li J V, Hadfield S T, Hole S, Nicholson J K, Wilson I D, Howe P W A, Stanley P D, Holmes E: Application of NMR-based metabolomics to the investigation of salt stress in maize (*Zea mays*). Phytochemical Analysis. 2011;**22**:214–224.

[28] Sato S, Arita M, Soga T, Nishioka T, Tomita M: Time-resolved metabolomics reveals metabolic modulation in rice foliage. BMC System Biology. 2008;**2**:51.

[29] Kusano M, Fukushima A, Kobayashi M, Hayashi N, Jonsson P, Moritz T, Ebana K, Saito K: Application of a metabolomic method combining one-dimensional and two-dimensional gas chromatography-time-of-flight/mass spectrometry to metabolic phenotyping of natural variants in rice. Journal of Chromatography B Analytical Technologies in the Biomedical and Life Sciences. 2007;**855**:71–79.

[30] Gong L, Chen W, Gao Y, Liu X, Zhang H, Xu C, Yu S, Zhang Q, Luo J: Genetic analysis of the metabolome exemplified using a rice population. Proceedings of the National Academy of Sciences of the United States of America. 2013;**110**:20320–20325.

[31] Zhou J, Ma C, Xu H, Yuan K, Lu X, Zhu Z, Wu Y, Xu G: Metabolic profiling of transgenic rice with cryIAc and sck genes: an evaluation of unintended effects at metabolic level by using GC-FID and GC-MS. Journal of Chromatography B Analytical Technologies in the Biomedical and Life Sciences. 2009;**877**:725–732.

[32] Roessner U, Wagner C, Kopka J, Trethewey R N, Willmitzer L: Simultaneous analysis of metabolites in potato tuber by gas chromatography-mass spectrometry. The Plant Journal. 2000;**23**:131–142.

[33] Carreno-Quintero N, Acharjee A, Maliepaard C, Bachem C W B, Mumm R, Bouwmeester H, Visser R G F, Keurentjes J J B: Untargeted metabolic quantitative trait loci analyses reveal a relationship between primary metabolism and potato tuber quality. Plant Physiology. 2012;**158**:1306–1318.

[34] Dobson G, Shepherd T, Verrall S R, Conner S, McNicol J W, Ramsay G, Shepherd L V T, Davies H V, Stewart D: Phytochemical diversity in tubers of potato cultivars and landraces using a GC-MS metabolomics approach. Journal of Agricultural and Food Chemistry. 2008;**56**:10280–10291.

[35] Dobson G, Shepherd T, Verall S R, Griffiths W D, Ramsay G, McNicol J W, Davies H V Stewart D: A metabolomics study of cultivated potato (*Solanum tuberosum*) groups andigena, phureja, stenotomum, and tuberosum using gas chromatography-mass spectrometry. Journal of Agricultural and Food Chemistry. 2010;**58**:1214–1223.

[36] Beckmann M, Enot D P, Overy D P, Draper J: Representation, comparison, and interpretation of metabolome fingerprint data for total composition analysis and quality trait investigation in potato cultivars. Journal of Agricultural and Food Chemistry. 2007;**55**:3444–3451.

[37] Catchpole G S, Beckmann M, Enot D P, Mondhe M, Zywicki B, Taylor J, Hardy N, Smith A, King R D, Kell D B, Fiehn O, Draper J: Hierarchical metabolomics demonstrates substantial compositional similarity between genetically modified and conventional potato crops. Proceedings of the National Academy of Sciences of the United States of America. 2005;**102**:14458–14462.

[38] Roessner U, Luedemann A, Brust D, Fiehn O, Linke T, Willmitzer L, Fernie A R: Metabolic profiling allows comprehensive phenotyping of genetically or environmental modified plant systems. Plant Cell. 2001;**13**:11–29.

[39] Defernez M, Gunning Y M, Parr A J, Shepherd L V T, Davies H V, Colquhoun I J: NMR and HPLC-UV profiling of potatoes with genetic modifications to metabolic pathways. Journal of Agricultural and Food Chemistry. 2004;**52**:6075–6085.

[40] Aliferis K A, Jabali S: FT-ICR/Ms and GC-EI/MS metabolomics networking unravels global potato sprout's responses to Rhizoctonia solani infection. PLoS One. 2012;**7**:e42576. doi:10.1371/journal.pone.0042576

[41] Schauer N, Zamir D, Fernie A R: Metabolic profiling of leaves and fruit of wild species tomato: a survey of the *Solanum lycopersicum* complex. Journal of Experimental Botany. 2005;**56**:297–307.

[42] Schauer N, Semel Y, Roessner U, Gur A, Balbo I, Carrari F, Pleban T, Pérez-Melis A, Bruedigam C, Kopka J, Willmitzer L, Zamir D, Fernie A R: Comprehensive metabolic profiling and phenotyping of interspecific introgression lines for tomato improvement. Nature Biotechnology. 2006;**24**:447–454.

[43] Tikunov Y, Lommen A, de Vos C H R, Verhoeven H A, Bino R J, Hall R D, Bovy A G. A: Novel approach for nontargeted data analysis for metabolomics large-scale profiling in tomato fruit volatiles. Plant Physiology. 2005;**139**:1125–1137.

[44] Mintz-Oron S, Mandel T, Rogachev I, Feldberg L, Lotan O, Yativ M, Wang Z, Jetter R, Venger I, Adato A, Aharoni A: Gene expression and metabolism in tomato fruit surface tissues. Plant Physiology. 2008;**147**:823–851.

[45] Le Gall G, Colquhoun I J, Davis A L, Collins G J, Verhoeyen V E: Metabolite profiling of tomato (*Lycopersicum esculentum*) using ^{1}H-NMR spectroscopy as a tool to detect potential unintended effects following a genetic modification. Journal of Agricultural and Food Chemistry. 2003;**51**:2447–2456.

[46] Moco S, Bino R J, Vorst O, Verhoeven H A, Groot J, van Beek T A, Vervoort J, deVos C H R: A liquid chromatography-mass spectrometry-based metabolome database for tomate. Plant Physiology. 2006;**141**:1205–1218.

[47] Moco S, Forshed J, deVos R C H, Bino R J, Vervoort J: Intra- and inter-metabolite correlation spectrocospy of tomato metabolism data obtained by liquid chromatography-mass spectrometry and nuclear magnetic resonance. Metabolomics. 2008;**4**:202–215.

[48] Fraser P D, Enfissi E M, Goodfellow M, Eguchi T, Bramley P M: Metabolite profiling of plant carotenoids using the matrix-assisted laser desorption ionization time-of-flight mass spectrometry. The Plant Journal. 2007;**49**:552–564.

[49] Bénard C, Bernillon S, Biais B, Osorio S, Maucourt M, Ballias P, Deborde C, Colombié S, Cabasson C, Jacob D, Vercambre G, Gautier H, Rolin D, Génard M, Fernie A R, Gibon Y, Annick Moing: Metabolomic profiling in tomato reveals diel compositional changes in fruit affected by source-sink relationship. Journal of Experimental Botany. 2015;**11**:3391–3404.

[50] Camanes G, Scalschi L, Vicedo B, González-Bosch C, García-Agustín P. An untargeted global metabolomic anlaysis reveals the biochemical changes underlying basal resistance and priming in Solanum lycopersicum, and identifies 1-methyltryptophan as a metabolite involved in plant responses to *Botrytis cinerea* and *Pseudomonas syringae*. The Plant Journal 2015;**84**:125–139.

[51] Bovy A, Schijlen E, Hall R D: Metabolic engineering of flavonoids in tomato (*Solanum lycopersicum*): the potential for metabolomics. Metabolomics. 2007;**3**:399–412.

[52] Pérez-Fons J, Wells T, Corol D I, Ward J L, Gerrish C, Beale M H, Seymour G B, Bramley P M, Frazer P D: A genome-wide metabolomic resource for tomato fruit from *Solanum pennellii*. Scientific Reports. 2014;**4**:1–8.

[53] Rodríguez-López C E, Hernández-Brenes C, Díaz de la Garza R I: A targeted metabolomics approach to characterize acetogenin profiles in avocado fruit (*Persea americana* Mill.). Royal Society of Chemistry Advances. 2015;**5**:106019–106029.

[54] Blakey R J, Tesfay S Z, Bertling I, Bower J P: Changes in sugars, total protein, and oil in 'Hass' avocado (*Persea Americana* Mill.) fruit during ripening. Journal of Horticultural Science & Biotechnology. 2012;**87**:381–387.

[55] Esteve C, D'Amato A, Marina M L, García M C, Righetti P G: Identification of avocado (*Persea Americana*) pulp proteins by nano-LC-MS/MS via combinatorial peptide ligand libraries. Electrophoresis. 2012;**33**:2799–2805.

[56] Hurtado-Fernández E, Pacchiarotta T, Mayboroda O A, Fernández-Gutiérrez A, Carrasco-Pancorbo A: Metabolomic analysis of avocado fruits by GC-APCI-TOF MS: effects of ripening degrees and fruit varieties. Analytical and Bioanalytical Chemistry. 2015;**407**:547–555.

[57] Jiang H, Sólyom A, Timmermann B N, Gang D R: Characterization of gingerol-related compounds in ginger rhizome (*Zingiber officinale* Rosc.) by high-performance liquid chromatography/electrospray ionization mass spectrometry. Rapid Communications in Mass Spectrometry. 2005;**19**:2957–2964.

[58] Jian H, Timmermann B N, Gang D R: Use of liquid chromatography-electrospray ionization tandem mass spectrometry to identify diarylheptanoids in turmeric (*Curcuma longa* L.) rhizome. Journal of Chromatography A. 2006;**1111**:21–31.

[59] Iijima Y, Gang D R, Lewinsohn E, Pichersky E: Characterization of geraniol synthase from the peltate glands of sweer basil (*Ocimum basilicum*). Plant Physiology. 2004;**134**:370–379.

[60] Welham T, Domoney C: Temporal and spatial activity of a promoter from a pea enzyme inhibitor gene and its exploitation for seed quality improvement. Plant Science. 2000;**159**:289–299.

[61] Charlton A, Allnutt T, Holmes S, Chisholm J, Bean S, Ellis N, Mullineaux, Oehlschlager S P: NMR profiling of transgenic peas. Plant Biotechnology Journal. 2004;**2**:27–35.

[62] Widodo, Patterson J H, Newbigin E, Tester M, Bacic A, Roessner U: Metabolic responses to salt stress of barley (*Hordeum vulgare* L.) cultivars, Sahara and Clipper, which differ in salinity tolerance. Journal of Experimental Botany. 2009;**60**:4089–4103.

[63] Graham S, Amigues E, Migaud M, Browne R A: Application of NMR based metabolomics for mapping metabolite variation in European wheat. Metabolomics. 2009;**5**:302–306.

[64] Achnine L, Huhman D V, Farag M A, Sumner L W, Blount J W, R A Dixon: Genomics-based selection and functional characterization of triterpene glycosyltransferases from the model legume *Medicago truncatula*. The Plant Journal. 2005;**41**:875–887.

[65] Farag M A, Deavours B E, A. de Fátima, Naoumkina M, Dixon R A Sumner: Integrated metabolite and transcript profiling identify a biosynthetic mechanism for hispidol in *Medicago truncatula* cell cultures. Plant Physiology. 2009;**151**:1096–1113.

[66] Na Jom K, Frank T, Engel K H: A metabolite profiling approach to follow the sprouting process of mung beans (*Vigna radiata*). Metabolomics. 2011;**7**:102–117.

[67] Goossens A, Häkkinen S T, Seppänen-Laakso T, Biondi S, V de Sutter, Lammertyn F, Nuutila A M, Söderlund H, Zabeau M, Inzé D, Oksman-Caldentey K M: A functional genomics approach toward the understanding of secondary metabolism in plant cells. Proceedings of the National Academy of Sciences of the United States of America. 2003;**100**:8595–8600.

[68] Wahyuni Y, Ballester A R, Sudarmonowati E, Bino R J, Bovy A G: Metabolite biodiversity in pepper (*Capsicum*) fruits of thirty-two diverse accessions: variation in health-related compounds and implications for breeding. Phytochemisty. 2011;**72**:1358–1370.

[69] Pereira G E, Gaudillere J P, van Leeuwen C, Hilbert G, Lavialle O, Maucourt M, Deborde C, Moing A, Rolin D: ^{1}H-NMR and chemometrics to characterize mature grape berries in four wine-growing areas in Bordeaux, France. Journal of Agricultural and Food Chemistry. 2005;**53**:6382–6389.

[70] Stewart D, McDougall G J, Sungurtas J, Verrall S, Graham J, Martinusen I: Metabolomic approach to identifying bioactive compounds in berries: advances toward fruit nutritional enhancement. Molecular Nutrition & Food Research. 2007;**51**:645–651.

[71] Aharoni, A., C. H. R. de Vos, H. A. Verhoeven, C. A. Maliepaard, G. Kruppa, R. Bino, and D. B. Goodenowe: Nontargeted metabolome analysis by use of Fourier Transform Ion Cyclotron Mass Spectrometry. OMICS. 2002;**6**: 217–234.

[72] Hanhineva K, Rogachev I, Kokko H, Mintz-Oron S, Venger I, Käremlampi S, Aharoni A: Non-targeted analysis of spatial metabolite composition in strawberry (Fragaria × annanassa) flowers. Phytochemistry. 2008;**69**:2463–2481.

[73] Biais B, Beauvoit B, Allwood J W, Deborde C, Maucourt M, Goodacre R, Rolin D, Moing A: Metabolic acclimation to hypoxia revealed by metabolite gradients in melon fruit. Journal of Plant Physiology. 2010;**167**:242–245.

[74] Heuberger A L, Robison F M, Lyons S M A, Broeckling C D, Prenni J E: Evaluating plant immunity using mass spectrometry-based metabolomics workflows. Frontiers in Plant Science. 2014;**5**:1–11.

[75] Allwood J W, Ellis D I, Heald J K, Goodacre R, Mur L A: Metabolomic approaches reveal that phosphatidic and phosphatidyl glycerol phospholipids are major discriminatory non-polar metabolites in responses by *Brachypodium distachyon* to challenge by *Magnaporthe grisea*. The Plant Journal. 2006;**46**:351–368. doi:10.1111/j.1365-313X.2006.02692.x

[76] Wojakowska A, Muth D, Narozna D, Madrzak C, Stobiecki M, Kachlicki P: Changes of phenolic secondary metabolic profiles in the reaction of narrow leaf lupin (*Lupinus angustifolius*) plants to infections with *Colletotrichum lupini* fungus or treatment with its toxins. Metabolomics. 2013;**9**:575–589.

[77] Peluffo L, Lia V, Troglia C, Maringolo C, Norma P, Escande A, Esteban H H, Lytovchenko A, Fernie A R, Heinz R, Carrari F: Metabolic profiles of sunflower genotypes with contrasting response to *Sclerotinia sclerotiorum* infection. Phytochemistry. 2010;**71**:70–80.

[78] Choi Y H, Kim H K, Linthorst H J M, Hollander J G, Lefeber A W M, Erkelens C, Nuzillard J M, Verpoorte R: NMR metabolomics to revisit the tobacco mosaic virus infection in *Nicotiana tabacum* leaves. Journal of Natural Products. 2006;**69**:742–748.

[79] Leiss K A, Choi Y H, Verpoorte R, Klinkhamer P G L: An overview of NMR-based metabolomics to identify secondary plant compounds involved in host plant resistance. Phytochemistry Reviews. 2011;**10**:205–216.

[80] Holliwood K, Brison D R, Goodacre R: Metabolomics: current trends technologies and future trends. Proteomics. 2006;**6**:4716–4723.

[81] Armitage E G, Barbas C: Metabolomics in cancer biomarker discovery: current trends and future perspectives. Journal of Pharmaceutical and Biomedical Analysis. 2014;**87**:1–11.

[82] Putri S P, Nakayama Y, Matsuda F, Uchikata T, Kobayashi S, Marsubara A, Fujusaki E: Current metabolomics: practical applications. Journal of Bioscience and Bioengineering. 2013;**115**:579–589.

[83] Canela N, Rodríguez M A, Baiges I, Nadal P, Arola L: Foodomics imaging by mass spectrometry and magnetic resonance. Electrophoresis. 2016;**37**:1748–1767.

[84] Ashton O F O, Wong M, McGuie T K, Vather R, Wang Y, Requejo-Jackman C, Ramankutty P, Woolf A B: Pigments in avocado tissue and oil. Journal of Agricultural and Food Chemistry. 2006;**54:**10151–10158.

[85] Liao C, Peng Y, Ma W, Liu R, Li C, Li X: Proteomic analysis revealed nitrogen-mediated metabolic developmental, and hormonal regulation of maize (*Zea mays* L.) ear growth. Journal of Experimental Botany. 2012;**63**:5275–5288.

[86] Liu X, Robinson P W, Madore MA, Witney G W, Arpaia M L: "Hass" avocado carbohydrate fluctuation I. Growth and phenology. Journal of the American Society of Horticultural Science. 1999;**124:**671–675.

[87] Liu X, Sievert J, Arpaia M L, Madore M A: Postulated physiological roles of the seven-carbon sugars, mannoheptulose, and perseitol in avocado. Journal of the American Society for Horticultural Science. 2002;**127**:108–114.

[88] Meyer M D, Terry LA: Development of a rapid method for the sequential extraction and subsequent quantification of fatty acids and sugars from avocado mesocarp tissue. Journal of Agricultural and Food Chemistry. 2008;**56**:7439–7445.

[89] Morgan M, Osorio S, Gehl B, Baxter C J, Kruger N J, Ratcliffe R G, Fernie A R, Sweetlove L J: Metabolic engineering of tomato fruit organic acid content guided by biochemical analysis of an introgression line. Plant Physiology. 2013;**161**:397–407.

[90] Roesnnser-Tulani U, Hegemann B, Lytovchenko A, Carrari F, Bruedigam C, Granot D, Fernie A R: Overexpressing hexokinase reveals that the influence of hexose phosphorylation diminishes during fruit development. Plant Physiology. 2003;**133**:84–99.

[91] Tesfay S Z, Bertling I, Bower J P, Lovatt C: The quest for the function of "Hass" avocado carbohydrates: clues from fruit and seed development as well as seed germination. Australian Journal of Botany. 2012;**60**:79–86.

[92] Kim H K, Choi Y H, Verpoorte R. NMR-based metabolomic analysis of plants. Nature Protocols. 2010;**5**:536–549. doi:10.1038/nprot.2009.237

Section 3

Biomedical and Clinical Metabolomics

Chapter 9

Application of Metabolomics for the Diagnosis and Traditional Chinese Medicine Syndrome Differentiation of Chronic Heart Failure

Juan Wang, Jianxin Chen, Huihui Zhao and Wei Wang

Additional information is available at the end of the chapter

http://dx.doi.org/10.5772/66209

Abstract

Chronic heart failure (CHF) was characterized by the failure of enough blood supply from the heart to meet the body's metabolic demands, and the prevalence of CHF continuously increases globally. The personalized diagnosis of Traditional Chinese Medicine (TCM) classifies CHF into several different syndrome types, and integrating Western and TCM to treat CHF has proved a validated therapeutic approach. Over the last few years, there has been a rapidly growing number of metabolomics applications aimed at finding biomarkers that could assist diagnosis, provide therapy guidance, and evaluate response to therapy for individualized intervention of CHF. Thus, in this review, particular attention will be paid to the past successes in applications of state-of-the-art technology on metabolomics to contribute to biomarker discovery in CHF research.

Keywords: metabolomics, chronic heart failure, Chinese Medicine

1. Introduction

Chronic heart failure (CHF), a progressive clinical syndrome characterized by the inability of the heart to adequately pump blood to meet metabolic demands of the body, represents the final common ground of pathogenesis wherein various causes of heart damage converge [1].

Despite a substantial improvement in the survival rate after the onset of CHF due to increasing use of pharmacological interventions, mortality of patients suffering from CHF remains high. Since it has been well recognized that the incidence and prevalence of CHF are expected to further increase with the aging of the population, better strategy for the prevention and treatment of CHF patients is still needed.

1.1. Epidemiology of CHF

CHF, in recent years, has been a major cause of morbidity and mortality in the general population [2]. Both the increasing age of the population and success in the treatment of patients with acute myocardial infarction raise the prevalence and thus the economic expenditures of chronic heart failure [3]. Heart failure is not only a common, costly, disabling, and potentially fatal condition but also the leading cause of hospitalization in people older than 65 [4]. In developing countries, 2–3% of the population suffers from heart failure, but in people from 70 to 80 years old, it occurs in 20–30% [5]. In developed countries, around 2% of adults suffer from heart failure, but in people older than 65, this increases to 6–10% [6].

1.2. Personalized intervention of CHF based on Traditional Chinese Medicine (TCM) syndrome

With progress being made in bioinformatics and medical science, the view on health and disease in Western life sciences has shifted from standard protocol-based disease management to personalized medicine [7]. Based on personalized health and systematically diagnostic principles, Traditional Chinese Medicine (TCM) has proved effective to restore the self-regulatory ability of the human system by thousands of years in clinic. And using integrating TCM and Western medicine to treat CHF has been reported to enhance heart function and reduce related clinical symptoms, including expiratory dyspnea and chronic fatigue, and subsequently improve echocardiographic measures, 6-min walking distance test, and patients' quality of life [8]. TCM physicians also pay more attention to the overall maladjustments of functional status called "syndrome type" [9]. It is not simply an assemblage of diseases' signs and symptoms but also a functional status caused by the reaction to or interaction with environmental changes and pathogenic factors [10]. In other words, the essence of TCM "syndrome type" is disturbance in biological metabolism networks, the changes in concentration and relative proportions of metabolomic biomarkers resulting from the imbalance of the human system. For example, yin deficiency syndrome of CHF patients is described as low fever, night sweats, afternoon zygomaticus red, dysphoria (fever) in chest palms soles, dry mouth and throat, red tongue with little coating, and thready rapid pulse according to the Clinical Terminology of Traditional Chinese Medical Diagnosis and Treatment—Syndromes, and Yang deficiency syndrome of CHF is a cluster of symptoms including an aversion to coldness, dispirited feelings and lack of motivation, diarrhea before dawn, shortness of breath, frequent urination, edema, and liability to catch cold. Therefore, TCM syndrome, also defined as TCM pattern, is the essence of diagnosis and treatment in TCM.

2. Bringing metabolomics into the forefront of CHF research

The metabolome is the final downstream product of transcription and translation and is thus closest to the phenotype [11]. Dynamics of primary metabolism operate in timescales of seconds. These two characteristics allow the metabolome to be a sensitive and rapid measure of the system phenotype. Metabolomics were first defined in 1998 [12, 13]. Progress has been made in methodological technologies that have lead to the discovery of metabolomic biomarkers and greater knowledge regarding disease mechanism from that time. From the 1960s, applications of mass spectrometry (MS) [14] and nuclear magnetic resonance spectroscopy [15, 16] drove the first holistic studies of mammalian biofluids to be conducted. In the past years, the development of technology has given impetus to metabolomics to its current status. More than 13,000 studies searched in PubMed Database have proved metabolomics a routinely applied tool nowadays. However, metabolomics is still the younger and smaller sibling of proteomics, transcriptomics, and genomics. Metabolite profiles can provide a fingerprint of metabolic changes that characterize the mechanism of CHF, a progressive clinical syndrome, and also highlight the potential of metabolomic analysis in the evaluation of disease condition.

Metabolomics, as an important component of systematical biology, can be used to perform dynamic studies on noninjured tissues and organs in vivo and in vitro using noninvasive approaches under nearly physiological conditions. Therefore, metabolomic detection and analysis of biological samples may contribute to understand the biochemical changes associated with the progression of diseases. And identification of disease-associated metabolic biomarkers could allow early diagnosis of disease and establishment of predictive diagnostic systems. It is reported that metabolomics gives itself unparalleled advantage to the most common cardiovascular condition encountered in clinical practices, heart failure [17]. Metabolomics has also showed significant potential in TCM studies in recent years. And several studies [18, 19] combining metabolomic techniques and TCM syndrome types have demonstrated fingerprints of metabolic changes that characterize Western Medicine-diagnosed diseases, which highlighted the potential of metabolomics in the evaluation of disease condition and TCM-guided personalized treatment.

2.1. General procedures in which metabolomics can be used for diagnosis and biomarker discovery

Metabolomics operates with a workflow [20, 21] starting from a biological question and experiment, proceeding through sample collection and preparation, analytical experiment(s) to acquire data, data preprocessing and analysis followed by biological interpretation. The metabolomics experimental workflow involves the design of biological and analytical experiments, sample preparation, data acquisition, data preprocessing and analysis and data interpretation. This workflow leads to biological interpretation and reasoning, as shown in **Figure 1**.

1) Biological Experiment (Design of experiment; Sample collection)
2) Analytical Experiment (Sample preparation; Data acquisition)
3) Data Integration (Data pre-processing;Data analysis;Integration with metadata) **4) Analysis and metabolite identification**
5) Biological Interpretation (Metabolomic biomarkers;biological mechanism)

Figure 1. The general workflow of metabolomic experiment.

Specifically, analytes in a metabolomic sample comprise a highly complex mixture. Mass spectrometry (MS) is used to identify and to quantify metabolites after optional separation by liquid chromatography-mass spectrometry (LC-MS), gas chromatography-mass spectrometry (GC-MS), nuclear magnetic resonance (NMR) spectroscopy, and capillary electrophoresis (CE). The raw data usually consist of measurements performed on subjects under various conditions. These measurements may be digitized spectra, or a list of metabolite levels. The software, called XCMS, is one of the most widely cited mass spectrometry-based metabolomics software programs in scientific literature. Prior to multivariate data analysis, statistical analyses are performed using SIMCA-P+12 software (Umetrics, Umea, Sweden) as variables and then mean-centered and pareto-scaled. Principal component analysis (PCA) and orthogonal partial least-squares discriminant analysis (OPLS-DA) are commonly used for the processed data. Score and loading plots are calculated to demonstrate discriminatory metabolites for each group; clustering of samples with similar metabolic fingerprints can be detected. This clustering can elucidate patterns and assist in the determination of disease biomarkers.

Two generalized experimental strategies are applied, metabolic profiling (or metabolomics) and metabolite-targeted analysis. Metabolic profiling allows comprehensive phenotyping of genetically or environmentally modified systems. The study is devised so as to acquire data on a large scale of metabolites (100–1000s) followed by interrogation of these data to figure out biological differences. The design of metabolomic experiments is of great importance here as it becomes easier to introduce confounding factors that are not recognized during data analysis and which can falsify biological conclusions [22].

2.2. Advances in metabolomics techniques and related statistical methods

Without knowing which metabolites are of specific biological interest, powerful analytical methods are used to illustrate thousands of metabolites reproducibly in a single sample. Relative alterations in metabolite concentrations are researched, and at this stage, the absolute concentrations of metabolites are not generally identified. With the goal to figure out all metabolites, this is currently not technologically realized. A platform of sample preparation and analytical methods is applied to acquire good coverage of identified metabolites (e.g., give reference to the Husermet project where LC-MS, GC-MS, and NMR spectroscopy have all been used; www.husermet.org). The application of univariate and multivariate analysis tools [23–25] was performed to process data. These discovery-phase researches and hypothesis-generating or inductive studies [26], sometimes by the non-cognoscenti as a "fishing expedition," aim to detect new biological and metabolomic markers. Moreover, another strategy, driven from known biology where a limited number of metabolites (generally less than 20) are known to be biologically relevant before the biological experiment, is conducted, and accurate quantification of distinguished metabolites is analyzed in a designed approach. Under the remit of traditional analytical chemistry and biochemical assays commonly found in clinical laboratories, this strategy has been applied for decades.

3. Biomarkers and metabolomics studies on CHF

In studies of human beings, metabolomics has been applied to define biomarkers related to prognosis or diagnosis of a disease or drug toxicity/efficacy, and it is hoped to provide greater pathophysiological understanding of disease or therapeutic toxicity/efficacy by these studies [27]. Coupling bioinformatics and biostatistics with metabolomic technology platforms permits the identification and quantification of molecules to characterize the whole organism's response to diseases [28]. Studies have demonstrated that metabolomics lends itself ideally to the most common cardiovascular condition encountered in clinical practices, heart failure [29]. Serum metabolites collected from 52 patients developed with CHF and 57 controls were analyzed and 38 peaks illustrated significant differences between patients and controls. As the current gold-standard biomarker brain natriuretic peptide (BNP), two metabolites of pseudouridine including a modified nucleotide present in tRNA and rRNA, a marker of cell turnover, and the tricarboxylic acid cycle intermediate 2-oxoglutarate were at least as diagnostic metabolic biomarkers of heart failure [30]. Moreover, 2-hydroxy, 2-methylpropanoic acid, erythritol, and 2, 4, 6-trihydroxypyrimidine were also good discriminators for cases and controls observed. We identified metabolites from early experiments, also biomarkers in the future, and will need to lay a foundation for larger, prospective, externally validated researches in clinical studies. Our study used to apply a metabolomics approach to plasma obtained directly from patients in order to assess its accuracy and reliability in diagnosing CHF, which showed better performance in terms of both specificity and sensitivity. It should be noted that the heart is known to consume a diverse set of fuel substrates, including lactate, glucose, amino acids, ketones, and particularly free fatty acids (FFAs). The metabolites include energy metabolism-related molecules, lipid/protein complexes, and amino acids. Plasma samples from 39 CHF patients and 15 controls were analyzed by NMR spectroscopy. After processing the data, PCA and OPLS-DA were performed. The statistical model revealed good explained variance and predictability, and the diagnostic performance assessed by leave-one-out analysis exhibited 92.31% sensitivity and 86.67% specificity. The OPLS-DA score plots of spectra revealed good separation between case and control on the level of metabolites, and multiple biochemical changes indicated hyperlipidemia, alteration of energy metabolism, and other potential biological mechanisms underlying CHF. It was concluded that the NMR-based metabolomics approach demonstrated good performance to identify diagnostic plasma markers and provided new insights into metabolic process related to CHF.

It has become well recognized that alteration in energy metabolites is of great importance in chronic heart failure in recent years. Impairment in extraction of a wide range of metabolites, probably pointing to severe energy deficiency, always leads to cardiac dysfunction in CHF patients. Moreover, higher level of lactate and decreased glucose in plasma metabolites seem to aggravate the impairment of energetic pathway in patients with CHF. A few metabolites relevant only in distinguishing CHF patients from healthy controls could be associated with prolonged exertion. Therefore, we observed increased lactate during anaerobic exercise. When stores of glucose are low and concentration of oxaloacetate has been exhausted, acetone is generated accordingly. Furthermore, alanine is expected to rise resulting from gluconeogenesis when lactate is produced [31]. A rise in creatine, phosphorylated to phosphocreatine in muscle, may indicate a physiological state of energy depletion. Besides, glycolate is of great importance in energy generation by mitochondria. Pyruvate stem from glycolysis is diverted away from the pyruvate dehydrogenase and toward the lactate dehydrogenase reaction, which contributes to the increase in lactate. A rise in acetyl-CoA may facilitate the inhibition of pyruvate dehydrogenase, which continues to be stemmed from fatty acid oxidation; however, it accumulates on the account of lowered trichloroacetic acid (TCA) cycle flux. Stress on energy metabolism can also affect metabolic area in CHF patients. 3-Hydroxybutyrate, 2-hydroxyisobutyrate, and 3-methyladipate had proved significant in glucose and lipid metabolism by Lin's study [32]. Glycoprotein and carnitine are reported in oxidative metabolism in mitochondrial and hypoxemic stress [33]. These results suggested oxidative fuel decrease and a greater reliance on anaerobic metabolism of glucose for energy production in the plasma of CHF patients.

In CHF patients, increased low-density lipoprotein (LDL) and decreased high-density lipoprotein (HDL) were observed to be relative to lipolysis as a backup mechanism for energy generation. Previous proteomic analysis of left-atrial cardiomyocyte and tissue samples from the congestive heart failure model also found significant alterations in apolipoprotein levels [34]. As apolipoproteins play a significant role in lipid metabolism, the alterations in apolipoprotein concentrations indicate that lipid metabolic dysregulation may be relative to CHD. Other studies have shown a close relationship between CHF and lipid regulation, which may account for the high comorbidity between metabolic syndrome and CHF [35]. Along these lines, previous study examining FFA extraction in myocardial ischemia patients also has found decreased FFA extraction and oxidation [36]. Choline and its derivatives represent important constituents in phospholipid metabolism of cell membranes and have been previously identified as markers of cellular proliferation. To summarize, CHF patients have limited myocardial metabolic reserve and flexibility, which verify a preliminary hypothesis of association between lipid metabolic disorder and CHF.

The change of the TCA cycle may also be a metabolic marker in CHF patients. Impaired TCA cycle flux, derived from the catabolism of glucose and amino acids, appears to occur in part through limiting levels of anaplerotic substrates [37]. The fall in glutamate/glutamine uptake was observed in CHF patients because they would normally be transfinite to form the anaplerotic substrate α-ketoglutarate. An increase in alanine release was also noted. The net production of alanine likely occurs via transamination of pyruvate, with glutamate as the nitrogen donor in the alanine transaminase reaction. In this period, this metabolic signature is consistent with impaired glucose oxidation resulted in the diversion of pyruvate away from the TCA cycle and into alanine transaminase reactions. Some scholars hypothesize that CHF activates proteolysis of skeletal muscle and enhances branched amino acid oxidation, as was described in inflammatory states [38]. Besides that, increased proline had been reported to play a role in coronary atherosclerosis diseases [39]. In view of these studies, the heart's ability to maintain homeostasis via glycogeolysis, neoglucogenesis and ketogenesis is compromised with CHF, as well as altering the amino acids.

4. Diagnostic power of metabolomics in TCM syndromes of CHF

Despite advances in the drug treatment strategy for CHF, the number of deaths resulting from this condition continues to rise [40]. TCM pays special attention to the integrity and holism of the human body and its interrelationship with nature. Based on different symptoms and signs, TCM adheres to the basic principle to treat the same disease by different methods and different diseases by one method and emphasizes personalized treatment, which truly indicates the essence of TCM intervention [41]. Therefore, treatment based on syndrome differentiation is the core of TCM therapy or CHF. From the perspective of TCM, CHF may occur in all differentiation types, including qi deficiency and blood stasis, yang deficiency and water retention, yin deficiency, and so on. Many Chinese herbs have demonstrated safety and efficacy in the management of chronic heart failure in either animal models or humans [42, 43]. In addition, modern biologic research has entered an era of integrating various research technologies and methods to tackle difficult biological problems at biomolecular level as a whole, which is exemplified by studies in the new scientific fields of metabolomics. It is therefore crucial to investigate the potential correlation between TCM syndrome type and metabolites to develop novel therapeutic approaches for better treatment of CHF.

4.1. Qi deficiency and blood stasis syndrome

Based on TCM, we classified CHF patients into several syndromes. We investigated plasma metabolites of CHF patients with qi deficiency and blood stasis syndrome to illuminate new approaches to the diagnosis and identify metabolic signatures of TCM syndromes in CHF. Combining plasma metabolomics with TCM syndrome-type diagnosis showed the distinguished metabolites of CHF patients with qi deficiency and blood stasis syndrome, including energy metabolites (glucose, lactate, and glycoprotein), lipid/protein complexes [HDL, LDL/very low-density lipoprotein (VLDL)], and amino acids (alanine, glutamate, valine, glycine, proline, and carnitine). Therefore, this metabolomic method may demonstrate potential in understanding of TCM syndromes of CHF. It is indisputable that there are limitations to each study with any other new diagnostic approaches. Here, the effects of other confounding factors on the metabolic profiles, though, can be analyzed by further studies with large cohort required to validate this method. For plasma samples representing the effects of metabolism in different organs, it is also difficult to assign a metabolic fingerprint to specific metabolic processes [44]. However, it should be noted that the altered metabolites are reflected in CHF patients with certain TCM syndrome, which can be harnessed as markers of diseases.

Qi deficiency and blood stasis syndrome, as a major syndrome among CHF patients, show a distinct signature of altered metabolism, which includes increased level of lactate, gly-protein, low-density lipoprotein (LDL)/very low-density lipoprotein (VLDL), and lower levels of glucose, valine, proline, alanine, and carnitine. And glycoprotein is closely associated with the physiology and pathology of cells' growth and can affect human metabolic energy supply and cellular immunity [45]. Furthermore, increased level of LDL and VLDL was observed in CHF patients with qi deficiency and blood stasis syndrome, which were the most prominent factors differentiating from controls. This metabolomics profile could be associated with lipolysis as a backup mechanism for energy utilization, for apolipoproteins are of great weight in lipid metabolism. Meanwhile, the plasma levels of well-known essential and nonessential amino acids (such as alanine and valine) decreased in the CHF patients with qi deficiency and blood stasis syndrome, breaking the internal equilibrium of the body gradually. This is in line with Yan's research; a metabolomics study on the rat model of myocardial ischemia with this syndrome showed that increased inositol and decreased valine, glycine, and serine were closely associated with energy metabolism and oxidative stress response [46]. Besides that, carnitine, an important substance involved in fat metabolism and energy supply, reduced dramatically in these patients. Studies have confirmed that L-carnitine can increase the uptake of free fatty acids (FAA), which will make use of glucose as oxidative fuel in certain circumstances [47]. If carnitine was insufficient, the oxidation process in mitochondria will be affected, which leads to the imbalance of cell metabolism and heart diseases. Tricarboxylic acid cycle, carbohydrates, proteins, and fats would involve in the above metabolic processes, indicating a complicated metabolic disorder in CHF patients with qi deficiency and blood stasis syndrome. Therefore, this metabolomic method may demonstrate potential in understanding of qi deficiency and blood stasis syndrome of CHF. It is indisputable that there are limitations to each new diagnostic approach. The effects of other confounding factors on the metabolic profiles, though, can be analyzed by further studies with large cohort required to validate this method [47]. For plasma samples representing the effects of metabolism in different organs, it is also difficult to assign a metabolic fingerprint to specific metabolic processes. However, it should be noted that these altered metabolites are at least partly reflected in CHF patients with qi deficiency and blood stasis syndrome, which can be harnessed as markers of diseases for personalized treatment based on TCM. Further mechanistic studies regarding this issue are warranted.

4.2. Yang deficiency and water retention syndrome

Yang deficiency and water retention patients were observed to show higher levels of lactate, gly-protein, pyruvic acid, alanine, glutamate, and lower levels of glucose, low-density lipoprotein (LDL)/very low-density lipoprotein (VLDL), and high-density lipoprotein (HDL).

Based on TCM theory, yang deficiency is associated with signs of chronic, weak, hypofunction, hypometabolism, degenerative symptoms, and extremely common to be observed in late or severe stage of many diseases [48]. The metabolism pattern in CHF patients with yang deficiency and water retention syndrome demonstrated the decrease in glucose metabolism and the increase in lactate, alanine, and pyruvate, suggesting that the disorder of carbohydrate and energy metabolism in patients is more serious. There might be enhanced endogenous glucose production from gluconeogenesis and pyruvic acid change may indicate increased hepatic gluconeogenesis to provide extra pyruvate as a substrate for glucose [49]. That is why syndrome typically occurs in patients with chronic heart failure at stage III and stage IV. Moreover, increased glycoprotein indicates immune defects in the patients [50], while the generally lower lipoprotein levels including LDL/VLDL and HDL also suggest the insufficient absorption and utilization of protein in this phase. And higher excretion of measured metabolites (glutamate and alanine) could indicate that they might have more potential disturbance of renal function, and resultant missed metabolites that are necessary for carbohydrate and energy metabolism [51]. Previous study [52] investigated the urinary metabolites of syndrome in patients with chronic kidney disease and revealed that the key distinguished metabolites differed between yang deficiency and water retention syndrome and control group including alanine, diethylamine, proline, and so on. As essential substance in cellular activities, the deficiency will affect energy supply in all aspects of human body. Finally, these alterations are likely important contributing factors to the altered metabolite profiling of CHF patients with yang deficiency and water retention syndrome.

5. Conclusions and future perspectives

The future goals for metabolomics are the validation of existing biomarkers in terms of mechanism and translation to man, together with a focus on characterizing the individual health care. So far, metabolomics may be of special clinical relevance for the diagnosis of syndromes of CHF and uncovering metabolomic pathway and prognosis in some extent, which could lead to a better understanding and improvement of personalized interventions for CHF. Metabolomics has also shown great advantage to discover possible early biomarkers for the development of CHF and assess progression during treatment, which can aid the discovery of prognostic indicators of outcome and disease response to personalized therapy.

Acknowledgements

This work was supported by the National Department Public Benefit Research Foundation of China under grant no. 200807007 and the National Natural Science Foundation of China under grant no. 81302914.

Author details

Juan Wang, Jianxin Chen, Huihui Zhao and Wei Wang*

*Address all correspondence to: doctorjuanwang@163.com

Beijing University of Chinese Medicine, Beijing, China

References

[1] David L, Zsuzsanna H, Anna M, Ellamae S, Mayu S et al. (2011) Molecular signatures of end-stage heart failure. J Cardiac Fail 17:868–872.

[2] Nuria F, Emili V, Montse C, Montse B, Miguel C-A et al. (2016) Medical resource use and expenditure in patients with chronic heart failure: a population-based analysis of 88 195 patients. Eur J Heart Fail. 18(9):1132–40.

[3] Bundkirchen A, Schwinger RHG (2004) Epidemiology and economic burden of chronic heart failure. Eur Heart J. 6(suppl D), D57–D60.

[4] McMurray JJ, Pfeffer MA (2005) Heart failure. Lancet 365(9474):1877–1889.

[5] Krumholz HM, Chen YT, Wang Y, Vaccarino V, Radford MJ, et al. (2000) Predictors of readmission among elderly survivors of admission with heart failure. Am Heart J 139:72–77.

[6] Dickstein K, Cohen-Solal A, Filippatos G, MTCMurray JJV, Ponikowski P, et al. (2008) ESC Guidelines for the diagnosis and treatment of acute and chronic heart failure 2008 The Task Force for the Diagnosis and Treatment of Acute and Chronic Heart Failure 2008 of the European Society of Cardiology. Developed in collaboration with the Heart Failure Association of the ESC (HFA) and endorsed by the European Society of Intensive Care Medicine (ESITCM). Eur Heart J 29:2388–2442.

[7] Marcinkiewicz-Siemion M, Ciborowski M, Kretowski A, Musial WJ, Kaminski KA (2016) Metabolomics – A wide-open door to personalized treatment in chronic heart failure? Int J Cardiol 219:156–163.

[8] Yunlun L, Jianqing J, Chuanhua Y, Haiqiang J, Jingwen X et al. (2014) Oral Chinese herbal medicine for improvement of quality of life in patients with chronic heart failure: a systematic review and meta-analysis. Qual Life Res 23:1177–1192.

[9] Wang Z, Liu X, Ho RL, Lam CW, Chow MS (2016) Precision or personalized medicine for cancer chemotherapy: is there a role for herbal medicine. Molecules 21(7):889.

[10] Prasad S, Tyagi A (2015) Traditional medicine: the goldmine for modern drugs. Adv Tech. Biol. Med. 03(1).

[11] Goodacre R (2010) An overflow of what else but metabolism! Metabolomics 6:1–2.

[12] Oliver SG, Winson MK, Kell DB, Baganz F (1998) Systematic functional analysis of the yeast genome. Trends Biotechnol 16(9):373–378.

[13] Tweeddale H, Notley-McRobb L, Ferenci T (1998) Effect of slow growth on metabolism of Escherichia coli, as revealed by global metabolite pool ("Metabolome") analysis. J Bacteriol 180(19):5109–5116.

[14] Horning EC (1968) Use of combined gas-liquid chromatography and mass spectrometry for clinical problems. Clin Chem 14(8):777.

[15] Behar KL, Denhollander JA, Stromski ME, Ogino T, Shulman RG, Petroff OAC, Prichard JW (1983) High-resolution H-1 nuclear magnetic-resonance study of cerebral hypoxia in vivo. Proc Natl Acad Sci USA Biol Sci 80(16):4945–4948.

[16] Howells SL, Maxwell RJ, Peet AC, Griffiths JR (1992) An investigation of tumor H-1 nuclear-magnetic-resonance spectra by the application of chemometric techniques. Magn Reson Med 28(2):214–236.

[17] Wang J, Zhongfeng L, Jianxin C, Huihui Z, Liangtao L, Chan C, Xuegong X, Wenting Z, Kuo G, Bin L, Junpeng Z, Wei W (2013) Metabolomic identification of diagnostic plasma biomarkers in humans with chronic heart failure. Mol BioSyst 9:2618–2626.

[18] Van Wietmarschen H, Yuan K, Lu C, Gao P, Wang J, Xiao C, Yan X, Wang M, Schroën J, Lu A, Xu G, van der Greef J (2009) Systems biology guided by Chinese medicine reveals new markers for sub-typing rheumatoid arthritis patients. Clin Rheumatol 15:330–337.

[19] Wan JB, Bai X, Cai XJ, Rao Y, Wang YS, Wang YT (2013) Chemical differentiation of Da-Cheng-Qi-Tang, a Chinese medicine formula, prepared by traditional and modern decoction methods using UPLC/Q-TOFMS-based metabolomics approach. J Pharm Biomed Anal 83:34–42.

[20] Brown M, Dunn WB, Ellis DI, Goodacre R, Handl J, Knowles JD, O'Hagan S, Spasic I, Kell DB (2005) A metabolome pipeline: from concept to data to knowledge. Metabolomics 1(1):39–51.

[21] Dunn WB, Broadhurst D, Atherton HJ, Goodacre R, Griffin JL (2011) Systems level studies of Mammalian metabolomes: the roles of mass spectrometry and nuclear magnetic resonance spectroscopy. Chem Soc Rev. 40(1):387–426.

[22] Broadhurst DI, Kell DB (2006) Statistical strategies for avoiding false discoveries in metabolomics and related experiments. Metabolomics 2(4):171–196.

[23] Smilde AK, Westerhuis JA, Hoefsloot HCJ, Bijlsma S, Rubingh TCM, Vis DJ, Jellema RH, Pijl H, Roelfsema F, van der Greef J (2010) Dynamic metabolomic data analysis: a tutorial review. Metabolomics 6(1):3–17.

[24] Madsen R, Lundstedt T, Trygg J (2010) Chemometrics in metabolomics—a review in human disease diagnosis. Anal Chim Acta 659(1–2):23–33.

[25] Cantor GH (2011) Metabolomics and mechanisms: sometimes the fisher catches a big fish. Toxicol Sci. 118(2):321–3.

[26] Mamas M, Dunn WB, Neyses L, Goodacre R (2011) The role of metabolites and metabolomics in clinically applicable biomarkers of disease. Arch Toxicol 85:5–17.

[27] Sabatine MS, Liu E, Morrow DA, Heller E, McCarroll R, et al. (2005) Metabolomic identification of novel biomarkers of myocardial ischemia. Circulation 112:3868–3875.

[28] Dunn WB, Ellis DI (2005) Metabolomics: current analytical platforms and methodologies. Trac-Trend Anal Chem 24:285–294.

[29] Dunn WB, Broadhurst DI, Deepak SM, Buch MH, McDowell G, et al. (2007) Serum metabolomics reveals many novel metabolic markers of heart failure, including pseudouridine and 2-oxoglutarate. Metabolomics 3:413–426.

[30] MacIntyre D, Jimenez B, Lewintre EJ, Martín CR, Schäfer H, et al. (2010) Serum metabolome analysis by 1H-NMR reveals differences between chronic lymphocytic leukaemia molecular subgroups. Leukemia 24:788–797.

[31] Lin D, Hollander Z, Meredith A, Stadnick E, Sasaki M, et al. (2011) Molecular signatures of end-stage heart failure. J Card Fail 17:867–874.

[32] Kumps A, Duez P, Mardens Y (2002) Metabolic, nutritional, iatrogenic, and artifactual sources of urinary organic acids: a comprehensive table. Clin Chem 48:708–717.

[33] De Souza AI, Cardin S, Wait R, Chung YL, Vijayakumar M, et al. (2010) Proteomic and metabolomic analysis of atrial profibrillatory remodelling in congestive heart failure. J Mol Cell Cardiol 49:851–863.

[34] Pauly DF, Pepine CJ (2003) The role of carnitine in myocardial dysfunction. Am J Kidney Dis 41:S35–S43.

[35] Oka T, Itoi T, Terada N, Nakanishi H, Taguchi R, et al. (2008) Change in the membranous lipid composition accelerates lipid peroxidation in young rat hearts subjected to 2 weeks of hypoxia followed by hyperoxia. Circ J.72(8):1359–66.

[36] Turer AT, Stevens RD, Bain JR, Muehlbauer MJ, van der Westhuizen J, et al. (2009) Metabolomic profiling reveals distinct patterns of myocardial substrate use in humans with coronary artery disease or left ventricular dysfunction during surgical ischemia/reperfusion. Circ J 119:1736–1746.

[37] Russell R (1991) Changes in citric acid cycle flux and anaplerosis antedate the functional decline in isolated rat hearts utilizing acetoacetate. J Clin Invest 87:384.

[38] Zimmerli LU, Schiffer E, Zürbig P, Good DM, Kellmann M, et al. (2008) Urinary proteomic biomarkers in coronary artery disease. Mol Cell Proteomics 7:290–298.

[39] Desai AS, Claggett B, Pfeffer MA, Bello N, Finn PV, Granger CB, MTCMurray JJV, Pocock SS (2013) Influence of hospitalization for cardiovascular versus noncardiovascular reasons on subsequent mortality in patients with chronic heart failure across the spectrum of ejection fraction. Circulation Heart Journal, 34(suppl 1):284.

[40] Li X, Luo XG, Lu X Duan, JG, Xu GW (2011) Metabolomics study of diabetic retinopathy using gas chromatography –mass spectrometry: a comparison of stages and subtypes diagnosed by Western and Chinese medicine. J Ethnopharmacol Mol BioSyst 7:2228–2237.

[41] Li X, Zhang J, Huang J, Ma A, Yang J, Li W, Wu Z, Yao C, Zhang Y, Yao W, Zhang B, Gao R (2013) Efficacy and safety of Qili Qiangxin capsules for Chronic Heart Failure Study Group. A multicenter, randomized, double-blind, parallel-group, placebo-controlled study of the effects of Qili Qiangxin capsules in patients with chronic heart failure. J Am Coll Cardiol 62(12):1065–1072.

[42] Guo N, Yang D, Wang X, Dai J, Wang M, Lei Y (2014) Metabolomic study of chronic heart failure and effects of Chinese herbal decoction in rats. J Chromatogr A 1362:89–101.

[43] Kannel WB, Ho K, Thom T, Chan G (1994) Epidemiological features of cardiac failure. Brit Heart J 72:S3.

[44] Chowdhury P, Kehl D, Choudhary R, Maisel A (2013) The use of biomarkers in the patient with heart failure. Curr Cardiol Rep 15(6):372.

[45] Yan B, A J, Hao HP, Wang GJ, Zhu XX, Zha WB, Liu LS, Guan EZ, Zhang Y, Gu Sh Huang Q, Zheng YT (2009) Metabolomic phenotype and identification of "heart blood stasis obstruction pattern" and "qi and yin deficiency pattern" of myocardial ischemia rat models. Sci China C Life Sci. 52(11):1081–90.

[46] Wei H, Pasman W, Rubingh C, Wopereis S, Tienstra M, Schroen J, Wang M, Verheij E, van der Greef J (2012) Urine metabolomics combined with the personalized diagnosis guided by Chinese medicine reveals subtypes of pre-diabetes. Mol Biosyst 8(5):1482–1491.

[47] Wang X, Aihua Z, Hui S (2013) Power of metabolomics in diagnosis and biomarker discovery of hepatocellular carcinoma. Hepatology 5:2072–2077.

[48] Tan Y, Liu X, Lu C, He X, Li J, Xiao C, Jiang M, Yang J, Zhou K, Zhang Z, Zhang W, Lu A (2014) Metabolic profiling reveals therapeutic biomarkers of processed Aconitum Carmichaeli Debx in treating hydrocortisone induced Kidney-Yang deficiency syndrome rats. J Ethnopharmacol 152(3):585–593.

[49] Conno SCr, Hansen MK, Corner A, Smith RF, Ryan TE (2010) Integration of metabolomics and transcriptomics data to aid biomarker discovery in type 2 diabetes. Mol BioSyst 6:909–921.

[50] Tedeschi S, Pilotti E, Parenti E, Vicini V, Coghi P, Montanari A, Regolisti G, Fiaccadori E, Cabassi A (2012) Serum adipokine zinc alpha 2-glycoprotein and lipolysis in

cachectic and noncachectic heart failure patients: relationship with neurohormonal and inflammatory biomarkers. Metab: Clin Exp 61(1):37–42.

[51] Wang X, Aihua Z, Hui S (2015) Power of metabolomics in diagnosis and biomarker discovery of hepatocellular carcinoma. Hepatology 5:2072–2077.

[52] Dong FX, Huang D, He LG, Jia W (2008) Research on urine metabolomics in chronic kidney disease Ⅲ with kidney-yang deficiency. China J Trad Chin Med Pharm 12(23): 1110–1113.

Chapter 10

A Metabolomics Approach to Metabolic Diseases

Luis Aldámiz-Echevarría, Fernando Andrade,
Marta Llarena and Domingo González-Lamuño

Additional information is available at the end of the chapter

http://dx.doi.org/10.5772/65469

Abstract

Metabolomics, defined as the comprehensive analysis of compounds in a biological specimen, is an emerging technology that helps several pathologies to inform about new biomarkers. Metabolic diseases comprise a group of rare conditions that in total represent an important health problem. Historically, small numbers of metabolites have been used to diagnose complex metabolic diseases such as diabetes or metabolic diseases. Metabolomic methodology, due to the evolution of clinical chemistry technologies, could detect thousands of organic compounds. In this way, metabolomic approach gives information of metabolic pathways describing physiopathology that underlies disease, including the possibility of discovery of new markers that could be used to diagnose or check the efficacy of the treatments. Diabetes, classic inborn error of metabolism as methylmalonic aciduria, lysosomal diseases and rare optic neuropathy affecting adults are discussed in this chapter.

Keywords: metabolic diseases, diabetes, optic neuropathy, lysosomal diseases

1. Introduction

Rare metabolic diseases comprise a group of more than 7000 conditions that are particularly rare, but in total, they represent an important health problem. There is no single, internationally accepted definition of a rare metabolic disease; currently, it is a syndrome that occurs in one child per 200,000 births. In Europe, they are supposed to affect 7% of the population. Nowadays, as less than 500 metabolic diseases have available and effective treatments, new therapeutic solutions should be developed.

New approach and initiatives are necessary to advance research for patients suffering from an inborn error of metabolism because governments pay little attention due to their costs and low incidences. Suitable and quick diagnosis is basic to the patients, even if there is no treatment, since it reduces the overdiagnosis, followed by suitable care, can improve quality of life for these patients.

The inherited metabolic diseases are biochemical defects diagnosed routinely by neonatal screening programmes. This successful screening methodology continues to broaden and improve, and new biomarkers are added depending on the country.

Metabolomics is an "omics" science focused on dynamic biology ruled by organic metabolites. A metabolome is defined as the group of metabolites detected and playing in the same metabolic pathway in the normal functioning of the cell. Metabolomics complements other analytical sciences as proteomics or genomics. Physiological changes that are consequences of a special gene expression are characterised by a variation in the metabolic compounds. Thus, the metabolome is more sensitive than the transcriptome or the proteome. A small variation in protein expression can have a significant effect on the activity of the metabolic pathway and the concentration of the relevant metabolites.

In summary, the workflow for metabolomics applied to metabolic diseases starts with model design or patients selection (inclusion/exclusion criteria), sample preparation and separation-detection of metabolites by chromatography and mass spectrometry, respectively. Then, the obtained information has to be aligned and identified in order to carry out the statistics. The last step in this methodology is to obtain useful biomarkers for each inborn error of metabolism (**Figure 1**).

The aim of this chapter is to show the utility of metabolomics, applying to diabetes and inborn error of metabolism. The latter ones traditionally with difficult diagnosis and uncertain treatment are due to their low incidence. However, the metabolomic approach could give these kinds of diseases a tool to discover new and effective biomarkers.

1.1. Diabetes

Diabetes, caused by several interactions between genetic and environmental factors, is a frequent disorder related to mutations in several genes, with each individual gene accounting for 1% of disease risk. In case of type 2 diabetes, dysfunction of multiple organ systems, including impaired insulin action in the muscle and adipose tissue, defective hepatic glucose production and insulin resistance are caused by loss of beta-cell mass and function. The difficult challenge for the understanding of the molecular pathways is evident, but progress in this area may be aided by the recent advent of technologies for metabolomic approach including nuclear magnetic resonance or mass spectrometry.

Metabolomic could provide some advantages over other "omics" technologies in diabetes research: (1) The amount of metabolites or small molecules found and identified is less than the number of genes or proteins. So, the pathway interpretation should be easier and effective, as the results from genomics and proteomics. (2) Metabolomics is a tool for describing mechanisms of action and reverse effects of several treatments. However, metabolomics can be a science with limitations for technology misuse or data overinterpretation.

http://dx.doi.org/10.5772/65469

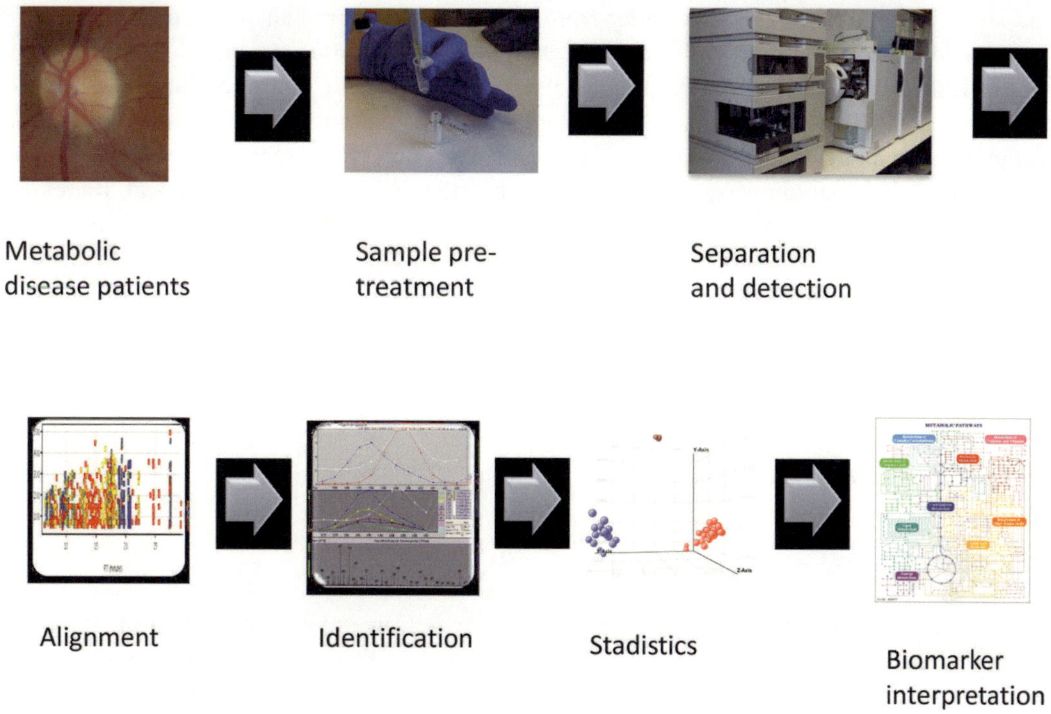

Figure 1. Experimental methodology involved in metabolomic studies of inborn errors of metabolism, where separation and detection phase could include a combination of gas chromatography, liquid chromatography and capillary electrophoresis with mass spectrometry.

Little information has been published about metabolic control of diabetic patients or their evolution with the treatment; for this reason, several profiles such as acylcarnitines, amino acids or cardiovascular markers, among them, should be studied taking into account recent metabolomic research for diabetes [1, 2]. Moreover, mitochondrial dysfunction and altered beta-oxidation of fatty acids have been described in diabetes [3].

A characteristic of all types of diabetes is that there are alterations in energy metabolism due to the lack of insulin action. This change is reflected especially in the metabolism of carnitine (**Figure 2**), which plays a major role as initiator of the beta-oxidation of fatty acids that occurs in the mitochondria. The investigation of the metabolic state of carnitine and its combination with the fatty acids and their metabolic intermediates produced during beta-oxidation seems essential to determine the degree of damage or the prognosis of these patients. Since it has been published that diabetic patients show carnitine deficiency and increased esterified carnitine [4], the study of the acylcarnitine composition compared to control group seems interesting, as well as the correlation of acylcarnitine composition with various complications which are expected during the course of the disease. These profiles of acylcarnitines can suggest incomplete beta-oxidation of long-chain fatty acids that can be amended with treatment. The carnitine and acylcarnitines can be analysed by means of their extraction from plasma or dried blood spots with an organic solvent, to remove proteins or solid-liquid extraction, respectively.

As some diabetic patients could suffer from alterations in energy metabolism, the study of creatine levels could be imperative to indicate a possible deficit of creatine to muscles or the brain. This creatine becomes creatine phosphate by a process of phosphorylation, leading to the genesis of ATP, essential in tissues with high-energy consumption, such as muscles and the brain through a process of dephosphorylation. An alteration of this metabolic pathway would lead to increase guanidinoacetate and a deficit in the production of creatine and, therefore, phosphocreatine.

Carnitine

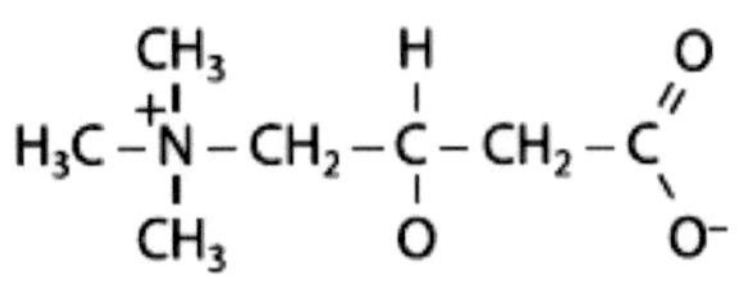

Figure 2. Molecular structure of carnitine, extracted from plasma by removing proteins.

In the absence of specific metabolic biomarkers of evolution and efficacy to treatment in diabetes, it should be performed a metabolic profile based on the profile of amino acids, which have been related to the methylation cycle and dementia (Met, Cys, Tau) in the diabetes.

L-Arginine, the main substrate of the endothelial nitric oxide synthase (NOS), is oxidised to L-citrulline and nitric oxide. Asymmetric dimethylarginine (ADMA) is derived from the proteolysis of proteins and acts as a competitive inhibitor of nitric oxide synthase, which is related to blood pressure, glucose intolerance, the thickness of the intima-media carotid, etc., making it a cardiovascular marker. It has been observed that the endothelial NO production is reduced in diabetic patients at high cardiovascular risk partly due to the limitation in the synthesis of arginine substrate and, secondly, because the ADMA, its inhibitor, is increased. So, it is proposed that the ADMA acts as a marker of cardiovascular risk in diabetic patients [5].

Epigenetics may contribute to metabolic study of diabetes through the global status of DNA methylation, which has been vaguely studied in this disease. Such mechanism methylation may play a role in the aetiology of monogenic diabetes. We propose the quantification of 5-methylcytosine (5-mC) in order to study changes in the methylation of DNA/RNA or its possible degradation during the treatment.

1.2. Methylmalonic aciduria

Methylmalonic aciduria (MMA) is an inherited disorder involving certain amino acid and fatty acid metabolism. Levels of methylmalonic acid are increased mainly in urine. They are caused by an enzymatic deficiency of methylmalonyl-CoA mutase (MUT) in the mitochondria or by a defect in the uptake, transport or synthesis of cobalamin, the cofactor of MUT. The management of this organic aciduria is based on a low-protein diet. MMA patient could suffer

http://dx.doi.org/10.5772/65469

from renal failure as one of their critical problems, not common in other organic acidurias. However, the cause of renal failure and cardiovascular complications depends on several pathways. So, it is thought that nitric oxide formation could be affected causing endothelial dysfunction. Related to nitric oxide, arginine-asymmetric dimethylarginine (ADMA) pathway, where ADMA acts as an inhibitor for nitric oxide synthase, should be described in these patients. ADMA levels are associated with an increase in cardiovascular events and chronic renal disease [6]. In addition, levels of the isomer symmetric dimethylarginine (SDMA) as a renal biomarker were also elevated in patients with chronic kidney disease.

Metabolic approach to evaluate the cardiovascular biomarker status should be recommended to improve the management for these patients. The objectives of this approach are also the identification and quantification of biomarkers of renal injury and inflammation such as interleukins, tumour necrosis factor alpha (TNFα) and transforming growth factor beta (TGFβ), in patients who are at different stages of kidney damage. The definition of biomarkers for early renal impairment in MMA patient could assess changes in these parameters during the kidney damage progression.

The comparison of MMA subgroups at different stages of kidney damage would allow the identification of those markers of inflammation and tissue fibrosis involved in tubulointerstitial nephritis or primary glomerular injury and will establish a chronology on the appearance of altered levels of biomarkers for the development of kidney damage. Finally, we believe that greater understanding of the role of these molecules and inflammatory processes in the context of associated nephropathy by means of metabolomics is vital to facilitate the development of novel targets and therapeutic nephroprotective strategies. The use of anti-inflammatory molecules or antioxidants can be applied clinically successfully.

1.3. Non-arteritic anterior ischemic optic neuropathy (NAION)

The cause for optic neuropathy affecting elderly patients could be multifactorial and difficult to diagnose in case of non-arteritic anterior ischemic optic neuropathy (NAION). Few information has been published about metabolic pathways involved in this rare disease. So, the purpose of metabolomics is to suggest effective biomarkers and to describe metabolic fingerprinting.

Samples of patients and controls can be fingerprinted with liquid chromatography coupled to high-sensitive mass spectrometry, such as quadrupole time of flight (QTOF). According to **Figure 1**, data should be filtered, aligned and statistically analysed before identifying new compounds. These biomarkers were found to be significant in class separation and could be later confirmed by obtaining the characteristic MS/MS spectra and online databases, as Metlin or Human Metabolome Database. NAION patients presented differences in the phospholipid profile in comparison with controls, such as lower levels of lysoPCs.

LysoPCs and lysoPEs are products or metabolites of PCs and PEs (**Figure 3**), respectively, which are structural components of cell membranes. The structure of lysoPCs is a choline polar group linked to fatty acyls that differ in chain length, position and degree of

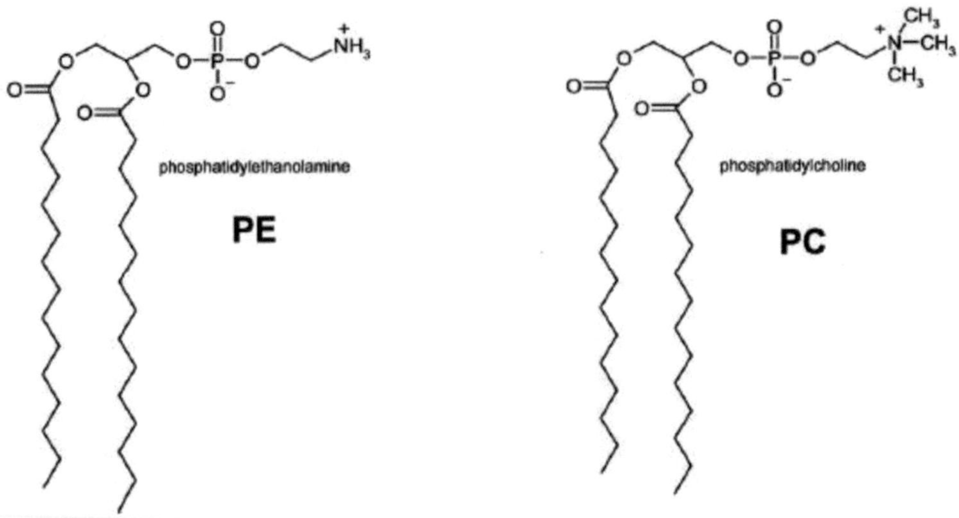

Figure 3. Molecular structure for phosphatidylcholine (PC) and phosphatidylethanolamine (PE), as the precursors or LysoPC and LysoPE.

saturation. These compounds were extracted from blood samples by centrifugation, so protein precipitation was performed by adding 300 μL of cool methanol:ethanol (1:1) to the plasma. They were vortexed during 1 min and stored at −20 °C for 10 min. After centrifugation, the supernatant can be injected in MS equipment. Increased levels of lysoPEs and reduced levels of lysoPCs can be a consequence of an altered choline phospholipid metabolism. The whole profile of lysoPCs has been published as a biomarker in other diseases, though the type of the lysoPCs was not taken into account [7]. As lysoPC profile is a biomarker for other diseases, variations in some lysoPC isomers may not be enough in order to identify a specific disease. For this reason, individual lysoPC levels produced by a pathology should be further investigated. As it has been researched by our group, lysoPC profiling of plasma shows a characteristic profile in NAION patients: higher levels of lysoPEs and lower levels of lysoPCs. As PCs are methylated products of PEs, the hypothesis for this result revolves around the poor methylation capacity in optic nerve cells. The characteristic profile of lysoPCs and lysoPEs also shows that the activity of phospholipases (PL) and lysophospholipases (LPL) could be increased [8].

NAION patients showed high levels of L-palmitoylcarnitine, which has the highest weight in the Partial Least Squares Discriminant Analysis (PLS-DA) prediction model (**Figure 4**). Nevertheless, these levels are not clearly related to the physiopathology of the disease because the uptake of acetylcarnitines by the mitochondria is a reversible process that is used to undergo mitochondrial β-oxidation of long-chain fatty acids and to transport acyl-coenzyme A from mitochondria to cytosol. So, more studies are necessary to confirm a disturbance in β-oxidation of fatty acid. However, these high levels of L-palmitoylcarnitine have been involved in the pathology of ischemia, acting as an inhibitor of cardiac Na,K-ATPase. Because of its amphipathic skill, palmitoylcarnitine can induce alterations in membrane fluidity and

http://dx.doi.org/10.5772/65469

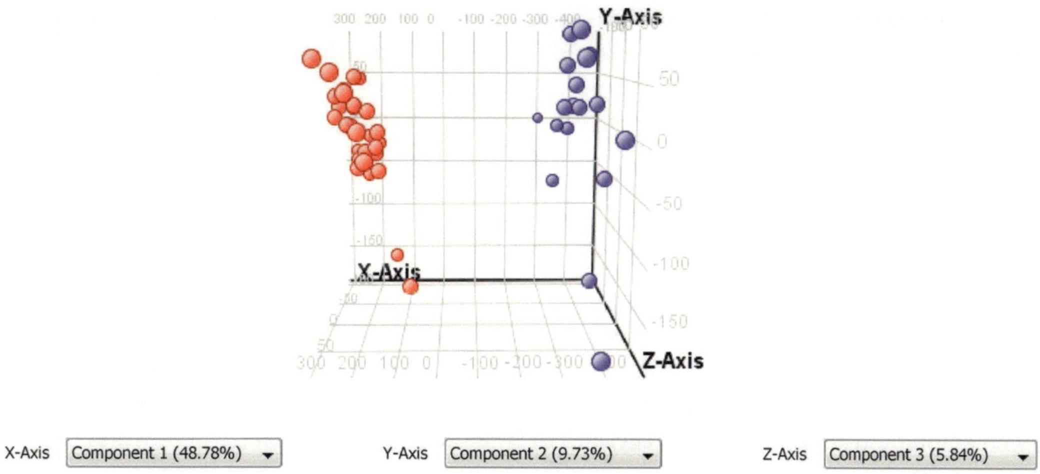

Figure 4. Partial Least Squares Discriminant Analysis (PLS-DA) scores plot of plasma metabolic profiles obtained from NAION patients and controls.

surface during apoptosis in NAION patients [9]. In this case, the membrane disturbance takes part in high-energy uptake cells of the optic nerve provoking severe blindness. This discussion could signify that carnitine as a supplement could increase the *in vivo* regeneration of optic nerve cells after NAION event. Furthermore, due to identified profile by metabolomics and NAION physiopathology as central nervous system problem, it seems to be a stronger relationship between NAION disease and other ischemic processes, such as myocardial ischemia or ictus, than with neuropathological or ophthalmological syndromes.

1.4. Lysosomal diseases

The lysosomal diseases are another group of inherited metabolic disorders characterised by mutations in genes encoding lysosomal enzymes and proteins. The enzyme deficiencies cause impaired intracellular turnover and build-up of complex molecules including sphingolipids, glycosaminoglycans and glycoproteins. These metabolites are established by metabolic approach as biomarkers. The pathology of lysosomal diseases is typically characterised by intra-lysosomal storage of a variety of substrates in multiple tissues and organs. Thus, the phenotypes of these disorders are complex and characterised by the variable association of visceral, skeletal and neurological manifestations.

The advancement in the diagnosis and treatment of lysosomal diseases has been largely stimulated by the improved knowledge on their molecular bases and pathophysiology. Metabolomics has a major role in this further impulse to the availability of technologies allowing faster diagnosis. In the case of Fabry disease, caused by deficiency of α-galactosidase A, the glycosphingolipid accumulation in plasma and urine is used as biomarker. Two specific biomarkers have been identified and quantified in plasma and urine: globotriaosylceramide (Gb3) and globotriaosylsphingosine (lyso-Gb3). The search continues for biomarkers that might be reliable indicators of disease severity and response to treatment.

Several metabolomic studies have been carried out for lysosomal diseases, such as Fabry [10] and Niemann-Pick [11] syndromes, to establish efficient biomarkers. For other lysosomal diseases like mucopolysaccharidoses, these studies could be useful to identify the increase of a specific glycosaminoglycan in urine, such as dermatan sulphate, chondroitin sulphate, heparan sulphate and keratan sulphate [12].

2. Future progress

Nowadays, the metabolomics is continuously improving its methodology to obtain new biomarkers quickly by means of advances in mass spectrometry and statistics software. So, it is clear that little-studied rare diseases, with no defined biomarkers, will be studied through this methodology. However, what is going to be an outstanding breakthrough is the combination of the information obtained by metabolomics, proteomics and genomics. Sometimes, these "omics" sciences are studied separately, but all information should be taken into account together in the future to understand the connexion of all information in the whole metabolic pathway.

3. Conclusions

It is evident that there is not a single perfect tool for global metabolic profiling. However, metabolomics, which combines sensitive analytical techniques with multivariate analysis, seems to be a good approach to identify biomarkers in diabetes and metabolic rare diseases, such as methylmalonic acidemia or lysosomal diseases.

Acknowledgements

Research and results in this manuscript were financed by the BioCruces and Carlos III Health Research Institutes.

Appendices and nomenclatures

ADMA	Asymmetric dimethylarginine
HMDB	Human Metabolome Database
LysoPC	Lysophosphatidylcholine
LysoPE	Lysophosphatidylethanolamine
MMA	Methylmalonic acidurias

NAION Non-arteritic anterior ischemic optic neuropathy

PC Phosphatidylcholine

PE Phosphatidylethanolamine

PLS-DA Partial least squares discriminant analysis

SDMA Symmetric dimethylarginine

Author details

Luis Aldámiz-Echevarría[1,*], Fernando Andrade[1], Marta Llarena[1] and Domingo González-Lamuño[2]

*Address all correspondence to: luisjose.aldamiz-echevarazuara@osakidetza.eus

1 Division of Metabolism, BioCruces Health Research Institute, Centre for Biomedical Research on Rare Diseases (CIBER-ER), Barakaldo, Bizkaia, Spain

2 Department of Nephrology and Metabolism, Marqués de Valdecilla University Hospital, Santander, Spain

References

[1] Bain JR, Stevens RD, Wenner BR, Ilkayeva O, Muoio DM, Newgard CB. Metabolomics applied to diabetes research: moving from information to knowledge. Diabetes 2009;58:2429-2443. doi: 10.2337/db09-0580

[2] Wang-Sattler R, Yu Z, Herder C, Messias AC, Floegel A, He Y, et al. Novel biomarkers for pre-diabetes identified by metabolomics. Mol Syst Biol 2012;8:615. doi: 10.1038/msb.2012.43

[3] Adams SH, Hoppel CL, Lok KH, Zhao L, Wong SW, Minkler PE, Hwang DH, Newman JW, Garvey WT. Plasma acylcarnitine profiles suggest incomplete long-chain fatty acid beta-oxidation and altered tricarboxylic acid cycle activity in type 2 diabetic African-American women. J Nutr 2009;139:1073-1081. doi: 10.3945/jn.108.103754

[4] Mamoulakis D, Galanakis E, Dionyssopoulou E, Evangeliou A, Sbyrakis S. Carnitine deficiency in children and adolescents with type 1 diabetes. J Diabetes Complications 2004;18:271-274. doi: 10.1016/S1056-8727(03)00091-6

[5] Tarnow L, Hovind P, Teerlink T, Stehouwer CD, Parving HH. Elevated plasma asymmetric dimethylarginine as a marker of cardiovascular morbidity in early diabetic nephropathy in type 1 diabetes. Diabetes Care 2004;27(3):765-769. doi: 10.2337/diacare.27.3.765

[6] Lu TM, Chung MY, Lin CC, Hsu CP, Lin SJ. Asymmetric dimethylarginine and clinical outcomes in chronic kidney disease. Clin J Am Soc Nephrol 2011;6:1566-1572. doi: 10.2215/CJN.08490910

[7] Dong J, Cai X, Zhao L, Xue X, Zou L, Zhang X, Liang X. Lysophosphatidylcholine profiling of plasma: discrimination of isomers and discovery of lung cancer biomarkers. Metabolomics 2010:6;478-488. doi: 10.1007/s11306-010-0215-x

[8] Ciborowski M, Rupérez FJ, Martinez-Alcázar MP, Angulo S, Radziwon P, Olszanski R, Kioczko J, Barbas C. Metabolomic approach with LC-MS reveals significant effect of pressure on Diver's plasma. J Proteome Res 2010;9:4131-4137. doi: 10.1021/pr100331j

[9] Nalecz KA, Miecz D, Berezowski V, Cecchelli R. Carnitine: transport and physiological functions in the brain. Mol Aspects Med 2004;25:551-567. doi: 10.1016/j.mam.2004.06.001

[10] Boutin M, Auray-Blais C. Metabolomic discovery of novel urinary galabiosylceramide analogs as Fabry disease biomarkers. J Am Soc Mass Spectrom 2015;26(3):499-510. doi: 10.1007/s13361-014-1060-3

[11] Maekawa M, Shimada M, Ohno K, Togawa M, Nittono H, Iida T, Hofmann AF, Goto J, Yamaguchi H, Mano N. Focused metabolomics using liquid chromatography/electrospray ionization tandem mass spectrometry for analysis of urinary conjugated cholesterol metabolites from patients with Niemann-Pick disease type C and 3β-hydroxysteroid dehydrogenase deficiency. Ann Clin Biochem 2015;52:576-587. doi: 10.1177/0004563214568871

[12] Staples GA, Zaia J. Analysis of glycosaminoglycans using mass spectrometry. Curr Proteomics 2011;8:325-336. doi: 10.2174/157016411798220871

Chapter 11

Metabolomics in Neonatology

Mina H. Hanna and Patrick D. Brophy

Additional information is available at the end of the chapter

http://dx.doi.org/10.5772/66295

Abstract

Throughout recent decades, the incidence of preterm birth has risen worldwide, and although the majority of preterm neonates now survive infancy, many suffer from debilitating morbidities in the short term and/or increased disease risks in the long term. Traditional diagnostic biomarkers suffer from considerable confounders, limiting their use in the early identification of diseases. There is a need to develop novel biomarkers that can identify, in real time, the evolution of organ dysfunction in an early diagnostic, monitoring, and prognostic fashion. Use of "omics," particularly metabolomics, may provide valuable information regarding functional pathways underlying different pathologies and prediction of clinical outcomes. The emerging knowledge generated by the application of metabolomics in neonatology provides new insights that can help to identify markers of early diagnosis, disease progression, response to treatment, and new therapeutic targets. In this chapter, we review the current knowledge of different metabolomics technologies in neonatal-perinatal medicine, including biomarker discovery, defining as yet unrecognized biologic therapeutic targets, and linking of metabolomics to relevant standard indices and long-term outcomes.

Keywords: metabolomics, biomarkers, personalized healthcare

1. Introduction

"Omics" refers to the collective technologies used to explore the roles, relationships, and actions of the various types of molecules that make up the phenotype of an organism. Living systems complexity and adaptiveness can be read through self-organized highly interconnected networks whose interacting components are dynamically coordinated in hierarchical patterns. Systems biology is a scientific discipline that endeavors to quantify all of the molecular elements of a biological system to assess their interactions and to integrate that information into network models. Therefore, systems biology reflects the knowledge acquired by omics in a meaningful manner [1].

From genes to metabolites, the omics technologies have progressed significantly in the medical field over the last decade secondary to the remarkable advancement in laboratory methodologies and analytical tools. We discuss in this chapter the current knowledge of metabolomics technologies in neonatal-perinatal medicine including biomarker discovery, defining as yet unrecognized biologic therapeutic targets, and linking of metabolomics to relevant standard indices.

2. Metabolomics technologies

The two major methodologies applied in metabolomics are nuclear magnetic resonance (NMR) spectroscopy and mass spectrometry (MS). Both techniques can deliver high sensitivity, selectivity, and throughput data with high degree of reproducibility [2]. NMR spectroscopy is a quantitative, nondestructive, reproducible technique that provides detailed information on solution-state molecular structures, based on atom-centered nuclear interactions. The advantage of applying NMR is that it uses the magnetic properties of atomic nuclei, delivering simultaneous information on both the structure and molecular mobility of metabolites without the need for the preselection of analytical parameters or sample derivatization procedures. However, sensitivity is a limiting factor and often metabolite concentrations in the range of 1–10 μmol/L are required for detection and quantification by NMR [3]. Mass spectrometry analytical platforms tend to have much higher sensitivity, enabling extensive assessment of different metabolites in biological fluids or tissues samples. **Figure 1** illustrates the flow of the processes of metabolomics.

The first step is sample collection, consistency in collection and processing through standard operating procedures is important to avoid iatrogenic biases. Variables to consider in this step: (1) circadian variation and time of collection during the day, (2) nutritional impact, and (3) gestational age at birth and postnatal days of life. Following collection, samples may be stored for extended periods of time. However, metabolites stability over time should be a part of quality control measurements in conjunction with analytical variability. Prior to analysis, the samples have to be extracted into a suitable solvent using chromatography, commonly used method is either gas or liquid chromatography (GC or LC) followed by ionization in a fluid or matrix; and subsequently, metabolites are identified using a mass spectrometer on the basis of their mass-to-charge ratio (m/z) and their representation in the spectrum. Metabolite identification in MS is destructive based on fragmentation patterns either through the measurement of molecular mass (indicative of the molecular formula) or by collection of fragmentation mass spectra (indicative of molecular structure). Therefore, the application of this technology has the advantage of identifying novel metabolites not previously described in databases. On the other hand, ion suppression in complex biological samples limits the ability to quantify metabolites secondary to the interaction of multiple analytes that are present in the ionization source at the same time [4, 5].

Metabolome characterization can be performed in a targeted manner, or in a nontargeted (pattern-recognition) manner. The target method implicates identification and quantification of specific metabolites in a given biofluid or tissue extract by comparing the spectrum of

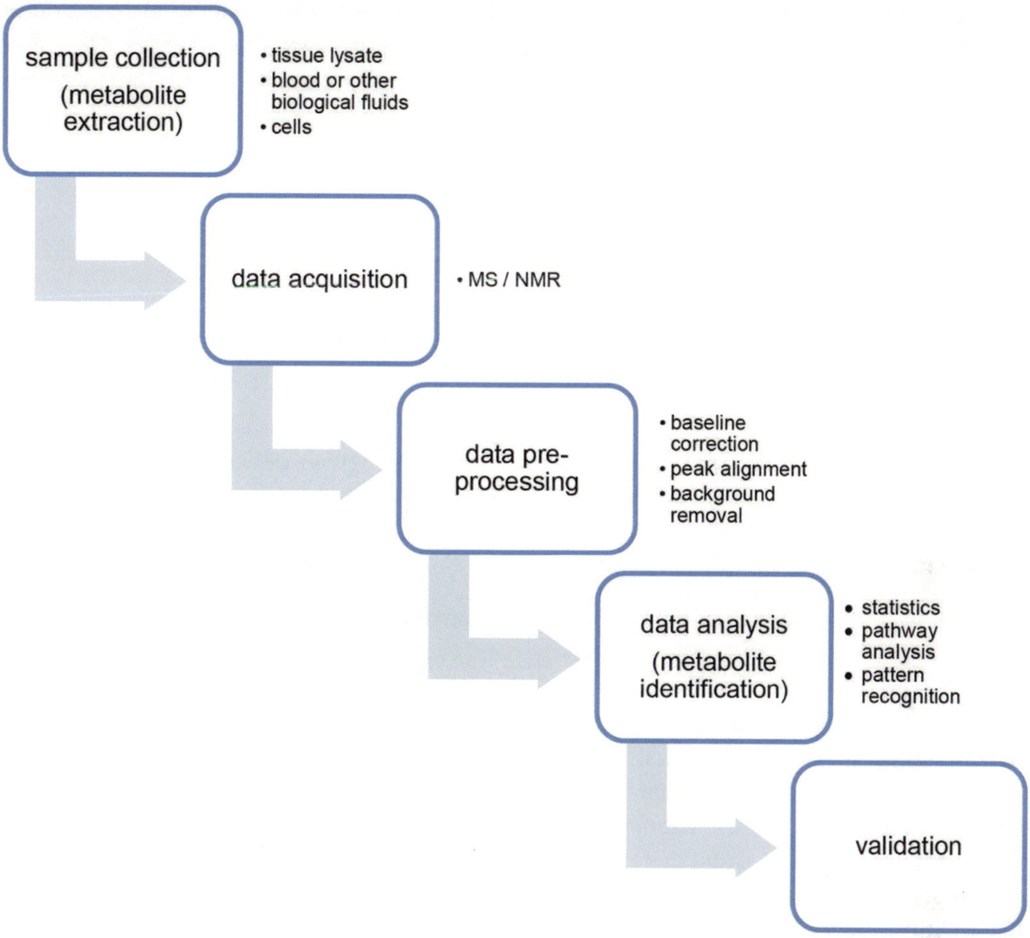

Figure 1. Metabolomics work flow.

interest to a library of reference spectra of pure compounds. This approach may suffer from an inherent bias as it captures only a part of the metabolome. Alternatively, the global nontargeted approach serves as a hypothesis-generating unbiased tool running as a first screening assay in clinical biomarker discovery studies, followed by targeted analysis for the metabolites that show significant differences or changes. The global pattern-recognition method can also screen for a multitude of key compounds in specific metabolic pathways which provide valuable information for metabolic fingerprinting.

The vast amount of data generated by metabolomics methods provides a unique opportunity to investigate alterations in metabolic pathways in response to changes in the cellular environment, and/or disease conditions. However, the high complexity of this data introduces a challenging aspect of data analysis that requires careful use of statistical methodologies and computational tools for efficient data visualization and analysis. Metabolic pathway analysis implies integration of the identified metabolites into metabolic correlation networks in order to better understand the complex relationships among various metabolites. Therefore, it allows researchers to correlate observed chemometric changes to the underlying pathological mechanisms.

3. Metabolomics in neonatology

3.1. Preterm birth and postnatal maturation

Preterm birth represents the aggregation of heterogeneous phenotypes, it is a complex disorder caused by multifactorial influences and the interplay of numerous risk factors.

Metabolomic profiling of amniotic fluid was able to distinguish patients who delivered at term from patients who delivered preterm. A decrease in carbohydrates was associated with preterm delivery in the presence or absence of inflammation whereas an increase in amino acid metabolites was a unique feature of preterm labor with inflammation [6].

Wilson et al. examined the associations between the degree of prematurity and the levels of amino acids, enzymes, and endocrine markers in a large cohort of infants. They concluded that children at different stages of prematurity are metabolically distinct [7]. Similarly Atzori et al. found that metabolomic analysis revealed distinct urinary metabolic profiles in neonates of different gestational ages, suggesting that gestational age has a strong effect on the metabolic profile of neonates, and applying this technology may predict the post-maturation of preterm and term neonates [8]. Furthermore, metabolomic analysis showed significant alterations in three metabolic pathways: (1) arginine and proline; (2) urea cycle; and (3) glycine, serine, and threonine between neonates with intrauterine growth restriction (IUGR) and controls [9].

3.2. Maternal chorioamnionitis and preeclampsia

The application of metabolomics methods has shown a clear distinction between preterm infants born to mothers with histological chorioamnionitis (HCA) from those born to mothers without HCA. Metabolites discriminating were the following: mannitol, 4-hydroxyphenylacetate, p-cresol, myo-inositol, trimethylamine-N-oxide, and 1-methylnicotinamide [10]. Similarly, metabolomics has the potential to identify changes under clinical conditions, such as preeclampsia (PE), that are associated with placental molecular pathophysiology. Heazell et al. have demonstrated that placental tissue from uncomplicated pregnancies cultured in 1% oxygen (hypoxia) had metabolic similarities to explants from preeclampsia pregnancies cultured at 6% oxygen (normoxia). This group of metabolites includes prostaglandins, a number of long-chain fatty acids and several amino acids [11]. Metabolic footprinting offers a hypothesis-generating strategy to investigate factors absorbed by and released from the placenta. Horgan et al. analyzed the metabolic footprint of placental villous explants cultured at different oxygen tensions between women who deliver a small for gestational age (SGA) baby and those from normal controls. SGA explant media cultured under hypoxic conditions was noted, on a univariate level, to exhibit the same metabolic signature as controls cultured under normoxic conditions for 49% of the metabolites of interest, suggesting that SGA tissue is acclimatized to hypoxic conditions *in vivo* [12].

3.3. Respiratory distress syndrome and bronchopulmonary dysplasia

Respiratory distress syndrome (RDS), formerly also known as hyaline membrane disease, is a common problem in preterm newborn infants. Surfactant deficiency or inactivation is a major contributing factor for the development of RDS. Metabolic profiling of bronchoalveolar lavage fluid (BALF) is a promising tool for assessing novel biomarkers of RDS in preterm infants. Applying GC-MS based metabolomic analysis revealed 10 metabolites that are over-expressed in BALF collected during mechanical ventilation following surfactant administration [13].

Bronchopulmonary dysplasia (BPD) is the most common chronic lung disease in infants with a multifactorial pathogenesis arising from a complex interaction between genetic and environment factors. Comparing the urinary metabolic profiles at birth of preterm neonates, Fanos et al. found five discriminant metabolites: lactate, taurine, trimethylamine-N-oxide (TMAO), myo-inositol (which increased in BPD patients), and gluconate (which was decreased) [14]. The increase in urinary lactate in the BPD group may represent a process of anaerobic respiration. Taurine and TMAO have anessential biological role for osmoregulation and membrane stabilization. Additionally, taurine has essential roles in calcium homeostasis, renal cell cycle and apoptosis, nerve cell activity and detoxification [15]. The data emerging from this study provide better insights into the pathophysiological mechanisms of BPD development.

3.4. Hypoxic ischemic encephalopathy

Hypoxic ischemic encephalopathy (HIE) is a complex neurological injury, characterized by biphasic depletion in high energy phosphates, with an estimated incidence of two per 1000 deliveries. Walsh et al. performed metabolomic analysis on umbilical cord blood from newborns that were divided into three groups: those with confirmed HIE (n = 31), asphyxiated infants without encephalopathy (n = 40) and matched controls (n = 71). Targeted metabolomic analysis showed a significant alteration between study groups in 29 metabolites from 3 distinct classes (amino acids, acylcarnitines, and glycerophospholipids). A logistic regression model using five metabolites clearly delineates severity of asphyxia andclassifies HIE infants with area under the curve (AUC) = 0.92 [16].

3.5. Necrotizing enterocolitis/late onset sepsis

Necrotizing enterocolitis (NEC) and late onset sepsis (LOS) are the leading causes of death among preterm infants. Stewart et al. compared the serum proteomic and metabolomic profiles longitudinally in preterm infants with NEC or LOS, matched to controls. While no single protein or metabolite was detectedin all NEC or LOS cases which was absent in controls; several proteins were identified which were associated with disease status. The expression of these proteins generally varied between diseased infants, potentially relating to differing pathophysiology ofdisease [17]. Similarly, Wilcock et al. found metabolomic differences in preterm babies at risk of NEC. However, sample sizes were insufficient to confidently identify a biomarker. Additionally, network modeling of preterm and term metabolomes suggested possible nutritional deficiency and altered pro-insulin action in preterm babies [18].

3.6. Neonatal kidney injury

Acute Kidney Injury (AKI) is common in neonates undergoing cardiac surgery, and is associated with increased mortality and ICU length of stay [19]. Mass spectrometry-based metabolomics was used in a prospective cohort of pediatric cardiac surgery patients ($n = 40$). Twenty-one of these children developed acute kidney injury defined as an increase in serum creatinine concentrations 50% or greater from baseline after 48–72 h. Homovanillic acid sulfate (HVA-SO_4), a dopamine metabolite was identified as a marker indicating AKI with 90% sensitivity and 95% specificity using a cutoff value of 24 ng/ml at 12 h after surgery [20]. Atzori et al. showed a correlation between urinary metabolic profiles and neutrophil gelatinase-associated lipocalin (NGAL) concentration in a cohort of young adults born with extremely low-birth weight (ELBW), using partial least-squares discriminant analysis [21].

3.6.1. Drug-induced nephrotoxicity

Nephrotoxic-medications are becoming increasingly recognized as a common and potentially modifiable cause of AKI in neonates. In a single center retrospective cohort 87% of very low birth weight infants (VLBW) were exposed to at least one nephrotoxic medication and on average these neonates were exposed to 14 days of nephrotoxic medications during their NICU stay [22]. Early identification of renal injury through omics technologies implicates defining different biomarkers that rely on the mechanisms of toxicity of each drug or drug class [23]. In our experimental study, gentamicin-induced acute kidney injury in newborn rats resulted in a distinct urinary metabolic profile characterized by glucosuria, phosphaturia, and aminoaciduria that preceded changes in serum creatinine. Additionally, lower levels of kynurenic acid were noted in the urine of gentamicin injected rats, coinciding with higher levels of tryptophan, suggesting a degrading effect of gentamicin toxicity on tryptophan metabolism pathway [24]. Xu et al. applied integrated pathway analysis and metabolite-transcript correlation analysis to define perturbed biochemical pathways and molecular functions that may be relevant to the mechanisms of nephrotoxicity. They concluded that transcriptional downregulation of luminal sodium-dependent transporters SLC5A1, SLC5A2, SLC6A18, and SLC16A7 might be the central mediators of drug-induced kidney injury and adaptive response pathways. The integrated pathway analysis performed on these studies indicates that cisplatin- or gentamicin-induced renal Fanconi-like syndromes manifested by glucosuria, hyperaminoaciduria, lactic aciduria, and ketonuria might be better explained by the reduction of functional proximal tubule transporters rather than by the perturbation of metabolic pathways inside kidney cells [25].

An alternative approach implicatesdiscoveryof a limited number of biomarkers that identify injury specific to primary sites in the kidney, such as the glomerulus or the proximal tubule. A prospective observational trial showed that the urinary excretion of biomarkers that signify proximal tubular damage was higher in the gentamicin group compared with control and preceded the peak of SCr and urine output decrease [26]. The application of different omics technologies *in vitro* systems and preclinical models to predict nephrotoxicity allows testing of the safety and efficacy of novel therapies and enhances the development and implementation of new drugs.

Askenazi et al. demonstrated that urinary biomarkers can predict AKI and mortality in very low birth weight infants independent of gestational age and birth weight [27]. We found that urinary NGAL, osteopontin(OPN) and cystatin C (Cys C) increased significantly in infants who developed AKI, in contrast, urinary epidermal growth factor (EGF) and uromodulin (UMOD) decreased significantly in this group. Urinary biomarkers demonstrated a significant change 24 h prior to contemporary creatinine-based neonatal AKI definition [28]. It is particularly important to recognize the differences in omics biomarkers across different gestational ages, postnatal days, and fluid balance status when designing future validation studies [29–32].

4. Future directions

The application of metabolomics approaches in neonatology is currently experimented on different platforms due to its unique ability to generate functional readouts of systems biology, setting the ground for future personalized prenatal, neonatal, and pediatric care. Yet the clinical translation of this unprecedented large amount of data into clinical practices for neonatal health care requires addressing of the inherent interindividual variability [33]. Metabolomics has the greatest potential in the field of biomarker discovery because this technique defines the signature of the actual processes that are occurring within the body rather than just merely examining compounds (such as untranscribed DNA or pre- or post-translationally modified proteins) that may be redundant to these processes. Although currently omics studies are mainly descriptive in nature, the goal is that through integration of experimental approaches and computational modelling, better models for personalized health care delivery will be generated. The following stages delineate how to translate the biomarker(s) discovery association studies into clinical applications in a stepwise approach:

1. Defining a clear clinical question
2. Selection of appropriate patient group(s), samples, and clinical data collection
3. Identification of specific biomarkers
4. Validation in a separate group of patients/samples
5. Validation on a large scale
6. Delineation of the interaction of nutrition, gene expression, and metabolism throughintegration of pathophysiology using pathway analysis tools

Experimental work in model systems and integration with other omics approaches are essential steps to provide insight into the pathophysiologic interactions between selected biomarkers and disease pathogenesis. Finally, large epidemiological cohort studies are needed to assess whether metabolomic biomarkers improve upon existing disease markers and to determine the validity of their application in different clinical settings.

5. Summary

The rapidly expanding field of metabolomics has been driven in recent years byadvances in the analytical methods. Metabolomics will have major implications in the field of personalized health care in the future. After establishing metabolomic profiles in the neonatal population, the next step is metabolic fingerprinting. In such metabolomic investigations, the intention is not to identify each observed compound but to compare patterns or fingerprints of metabolites that change in response to disease or drug exposure. The combination of metabolic profiling and fingerprinting will lead to the maximum utilization of metabolomics. In one approach, changes in fingerprints correlating with metabolite profiles may be linked to a physiological or pathological state. As more quantitative metabolomic databases evolve, they can be integrated with data sets from the other "omics" technologies to enhance the data value and provide greater biological insight than anyone "omics" technique alone can offer. The promise of this emerging technology is focusing on translational metabolomics for the identification of biomarkers, monitoring postnatal metabolic maturation, and the implementation of a tailored management of neonatal disorders.

Author details

Mina H. Hanna[1]* and Patrick D. Brophy[2]

*Address all correspondence to: mina.hanna@uky.edu

1 Department of Pediatrics, University of Kentucky, Lexington, KY, USA

2 Department of Pediatrics, University of Iowa, Iowa City, IA, USA

References

[1] Pesce F, Pathan S, Schena FP. From -omics to personalized medicine in nephrology: integration is the key. Nephrol Dial Transplant 2013;28:24–8.

[2] Lindon JC, Nicholson JK. Spectroscopic and statistical techniques for information recovery in metabonomics and metabolomics. Annu Rev Anal Chem (Palo Alto Calif) 2008;1:45–69.

[3] Schnackenberg LK, Beger RD. Monitoring the health to disease continuum with global metabolic profiling and systems biology. Pharmacogenomics 2006;7:1077–86.

[4] Pan Z, Raftery D. Comparing and combining NMR spectroscopy and mass spectrometry in metabolomics. Anal Bioanal Chem 2007;387:525–7.

[5] Feng X, Liu X, Luo Q, Liu BF. Mass spectrometry in systems biology: an overview. Mass Spectrom Rev 2008;27:635–60.

[6] Romero R, Mazaki-Tovi S, Vaisbuch E, et al. Metabolomics in premature labor: a novel approach to identify patients at risk for preterm delivery. J Matern Fetal Neonatal Med 2010;23:1344–59.

[7] Wilson K, Hawken S, Ducharme R, et al. Metabolomics of prematurity: analysis of patterns of amino acids, enzymes, and endocrine markers by categories of gestational age. Pediatr Res 2014;75:367–73.

[8] Atzori L, Antonucci R, Barberini L, et al. 1H NMR-based metabolomic analysis of urine from preterm and term neonates. Front Biosci (Elite Ed) 2011;3:1005–12.

[9] Dessi A, Atzori L, Noto A, et al. Metabolomics in newborns with intrauterine growth retardation (IUGR): urine reveals markers of metabolic syndrome. J Matern Fetal Neonatal Med 2011;24 Suppl 2:35–9.

[10] Syggelou A, Iacovidou N, Atzori L, Xanthos T, Fanos V. Metabolomics in the developing human being. Pediatr Clin North Am 2012;59:1039–58.

[11] Heazell AE, Brown M, Worton SA, Dunn WB. Review: the effects of oxygen on normal and pre-eclamptic placental tissue--insights from metabolomics. Placenta 2011;32 Suppl 2:S119–24.

[12] Horgan RP, Broadhurst DI, Dunn WB, et al. Changes in the metabolic footprint of placental explant-conditioned medium cultured in different oxygen tensions from placentas of small for gestational age and normal pregnancies. Placenta 2010;31:893–901.

[13] Fabiano A, Gazzolo D, Zimmermann LJ, et al. Metabolomic analysis of bronchoalveolar lavage fluid in preterm infants complicated by respiratory distress syndrome: preliminary results. J Matern Fetal Neonatal Med 2011;24 Suppl 2:55–8.

[14] Fanos V, Pintus MC, Lussu M, et al. Urinary metabolomics of bronchopulmonary dysplasia (BPD): preliminary data at birth suggest it is a congenital disease. J Matern Fetal Neonatal Med 2014;27 Suppl 2:39–45.

[15] Piersigilli F, Bhandari V. Biomarkers in neonatology: the new "omics" of bronchopulmonary dysplasia. J Matern Fetal Neonatal Med 2016;29:1758–64.

[16] Walsh BH, Broadhurst DI, Mandal R, et al. The metabolomic profile of umbilical cord blood in neonatal hypoxic ischaemic encephalopathy. PLoS One 2012;7:e50520.

[17] Stewart CJ, Nelson A, Treumann A, et al. Metabolomic and proteomic analysis of serum from preterm infants with necrotising entercolitis and late-onset sepsis. Pediatr Res 2016;79:425–31.

[18] Wilcock A, Begley P, Stevens A, Whatmore A, Victor S. The metabolomics of necrotising enterocolitis in preterm babies: an exploratory study. J Matern Fetal Neonatal Med 2016;29:758–62.

[19] Alabbas A, Campbell A, Skippen P, Human D, Matsell D, Mammen C. Epidemiology of cardiac surgery-associated acute kidney injury in neonates: a retrospective study. Pediatr Nephrol 2013;28:1127–34.

[20] Beger RD, Holland RD, Sun J, et al. Metabonomics of acute kidney injury in children after cardiac surgery. Pediatr Nephrol 2008;23:977–84.

[21] Atzori L, Mussap M, Noto A, et al. Clinical metabolomics and urinary NGAL for the early prediction of chronic kidney disease in healthy adults born ELBW. J Matern Fetal Neonatal Med 2011;24 Suppl 2:40–3.

[22] Rhone ET, Carmody JB, Swanson JR, Charlton JR. Nephrotoxic medication exposure in very low birth weight infants. J Matern Fetal Neonatal Med 2014;27:1485–90.

[23] Bonventre JV, Vaidya VS, Schmouder R, Feig P, Dieterle F. Next-generation biomarkers for detecting kidney toxicity. Nat Biotechnol 2010;28:436–40.

[24] Hanna MH, Segar JL, Teesch LM, Kasper DC, Schaefer FS, Brophy PD. Urinary metabolomic markers of aminoglycoside nephrotoxicity in newborn rats. Pediatr Res 2013;73:585–91.

[25] Xu EY, Perlina A, Vu H, et al. Integrated pathway analysis of rat urine metabolic profiles and kidney transcriptomic profiles to elucidate the systems toxicology of model nephrotoxicants. Chem Res Toxicol 2008;21:1548–61.

[26] Jansen D, Peters E, Heemskerk S, et al. Tubular injury biomarkers to detect gentamicin-induced acute kidney injury in the neonatal intensive care unit. Am J Perinatol 2016;33:180–7.

[27] Askenazi DJ, Montesanti A, Hunley H, et al. Urine biomarkers predict acute kidney injury and mortality in very low birth weight infants. J Pediatr 2011;159:907–12 e1.

[28] Hanna M, Brophy PD, Giannone PJ, Joshi MS, Bauer JA, RamachandraRao S. Early urinary biomarkers of acute kidney injury in preterm infants. Pediatr Res 2016;80(2):218–23.

[29] Saeidi B, Koralkar R, Griffin RL, Halloran B, Ambalavanan N, Askenazi DJ. Impact of gestational age, sex, and postnatal age on urine biomarkers in premature neonates. Pediatr Nephrol 2015;30:2037–44.

[30] Askenazi DJ, Koralkar R, Levitan EB, et al. Baseline values of candidate urine acute kidney injury biomarkers vary by gestational age in premature infants. Pediatr Res 2011;70:302–6.

[31] Askenazi D, Saeidi B, Koralkar R, Ambalavanan N, Griffin RL. Acute changes in fluid status affect the incidence, associative clinical outcomes, and urine biomarker performance in premature infants with acute kidney injury. Pediatr Nephrol 2016;31:843–51.

[32] Tsukahara H, Sugaya T, Hayakawa K, et al. Quantification of L-type fatty acid binding protein in the urine of preterm neonates. Early Hum Dev 2005;81:643–6.

[33] Hanna MH, Dalla Gassa A, Mayer G, et al. The nephrologist of tomorrow: towards a kidney-omic future. Pediatr Nephrol 2016.